Oxford Applied Mathematics and Computing Science Series

General Editors
J. Crank, H. G. Martin, D. M. Melluish

J. C. NEWBY
Brunel University

Mathematics for the biological sciences

FROM GRAPHS THROUGH CALCULUS
TO DIFFERENTIAL EQUATIONS

CLARENDON PRESS · OXFORD
1980

Oxford University Press, Walton Street, Oxford OX2 6DP

OXFORD LONDON GLASGOW
NEW YORK TORONTO MELBOURNE WELLINGTON
KUALA LUMPUR SINGAPORE JAKARTA HONG KONG TOKYO
DELHI BOMBAY CALCUTTA MADRAS KARACHI
NAIROBI DAR ES SALAAM CAPE TOWN

Published in the United States by Oxford University Press, New York

British Library Cataloguing in Publication Data

Newby, J C
    Mathematics for the biological sciences. —
    (Oxford applied mathematics and computing science
    series).
    1. Biomathematics
    I. Title   II. Series
    510'.2'4574   QH323.5   79-41390
    ISBN 0-19-859623-5
    ISBN 0-19-859624-3 Pbk

Typeset by Anne Joshua Associates, Oxford
and printed in Great Britain by Butler and Tanner Ltd, Frome

Preface

THE AUTHOR of any book written with the intention of helping the reader to learn something knows that it is unlikely to be read for relaxation. This is particularly so in the case of books on mathematics. If, in addition, the mathematics is in the context of another subject the problems facing both the author and reader are further compounded. At best the new undergraduate biologist may not have studied mathematics for some time and at worst he or she may have an active dislike of the subject. Surprise and possible dismay are experienced when it is realized that a course in mathematics is an integral part of the first-year biological studies.

It is with these attitudes in mind that this book has been written. I can hardly expect that the reader's feelings towards mathematics will undergo a dramatic change the minute he or she starts the first chapter. I have therefore assumed that the student reader is an apprehensive or even unwilling traveller in an unfamiliar environment and is probably only in this position because of the requirements of the undergraduate biology course. If this is the case the reader may want just sufficient mathematics to enable him or her to comply with these requirements. To facilitate this an informal approach is used and there are many worked examples to illustrate techniques and points arising in the text. The book may be used simply at this elementary level but it is to be hoped that such crude motivation will evolve into something more enlightened when it is realized that mathematics can be of use in biology not simply as a means of obtaining the 'right answer' but also as an aid to understanding some of the general principles encountered. To help with this process examples from biology and biochemistry are used throughout the book to illustrate the applicability of the mathematics. In addition an early chapter is devoted to the use of elementary mathematical methods in the design of experiments and the analysis of results.

It is probable that the majority of students reading this book will have had very little if any experience of using a text on mathematics without constant classroom supervision. For this reason a few words on how to use such a book may not come amiss. It should be obvious that a mathematics book requires careful reading and the student may have to adapt his or her reading technique to accommodate this. Each symbol has a meaning and in many cases indicates one or more instructions. The language of mathematics is of necessity very concise. It is a good idea to have available a pencil and paper on which to work through details that the reader may not follow in the text. Working through the examples and exercises is of great importance. This is not just a mechanical process which may be accomplished with a dormant mind. The exercises vary considerably in purpose and difficulty. Apart from gaining valuable experience by doing them the student will start to get a 'feel' for the mathematics involved and obtain insight which will make understanding and learning considerably easier. If trouble is experienced with a particular exercise try to locate the source. Is it failure to understand what the question requires or failure to do what is required? In the former case advice should be sought. In the latter reference to the preceding text may well resolve the problem. It is important that difficulties are not left unresolved. Mathematics is a logical structure which relies entirely for its advance on the soundness of the preceding work. If any of this is omitted or not understood the student may suddenly, at some later time, find it impossible to move onwards.

At one time the only equipment that a mathematician needed was something to write with and something to write on. Today there are all kinds of devices which can make life easier from tables to computers. In order to study the material in this book the reader will require pencil and paper, ruler and graph paper, and a set of mathematical tables (four or five figures). When selecting a set of tables try to ensure that apart from the usual logarithms, trigonometric functions, square roots, and reciprocals, they also include natural or Naperian logarithms, positive and negative exponentials (e^x and e^{-x}), and conversions from degrees to radians. If a calculator is available with all these facilities the tables can be dispensed with.

The advent of the digital computer has meant that some types of problem can be efficiently solved using numerical methods and in the case of more advanced problems this may be the only method of solution. Some simple numerical methods have therefore been included in the text by way of introduction to this subject. All of the associated examples and exercises can be performed using just pencil, paper, and tables or calculator but if computing facilities are available they could be used. For this reason some short programs written in a simple language known as BASIC have been included. These may be ignored if such facilities are not available.

Uxbridge   J.C.N.
June 1979

Contents

relationship for germination and temperature. A quadratic
model of sapling growth. Rate of growth of a bounded
population.

pendulum. Wading birds. Body heat loss. Mass,
length and surface area of a species.

1 Graphs

1.1. Histograms

HISTOGRAMS or 'block diagrams' are used to represent data associated with intervals. Some simple examples are monthly rainfall, hourly traffic flow, and the number of adults in the population in consecutive 3-inch height intervals. The intervals used are normally equal but need not necessarily be so. The area of each block is a measure of the data being recorded. If the intervals are all equal then the height of each block is also a measure of the data.

As an example consider a histogram of the number of cases of a virus infection reported each day in Figure 1.1. The intervals are all

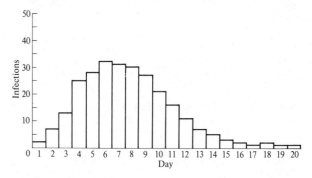

FIG. 1.1

equal and so the height of each column represents the daily infections. If however the records for days 6, 7, and 8 could not be distinguished and it was only known that a total of 93 cases were reported during this period and, similarly, that over days 9 and 10 a total of 48 cases were reported, then the intervals would not be equal. The histogram would now have to be drawn with a single column over days 6, 7, and 8 of height 31 so that its area would be 3 × 31 = 93. Similarly a single column of height 24 would be drawn over days 9 and 10 since 2 × 24 = 48 (see Figure 1.2).

It can be seen that the method of area representation preserves the

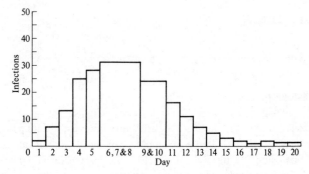

FIG. 1.2

features of the earlier more detailed histogram. In both cases it is clear that peak infection occurs after about a week and that further infections are unlikely after about three weeks.

Example. In a population of adult small mammals all were found to have masses between 20 and 40 g. The distribution was found to be as follows

Mass (g)	less than 25	25–6	26–7	27–8	28–9	29–30	30–1	31–2	32–3
No.	8	5	9	14	20	27	35	48	52

Mass (g)	33–4	34–5	35–6	36–7	greater than 37
No.	42	25	10	4	6

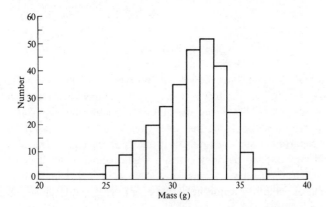

FIG. 1.3

Since all masses lie between 20 and 40 g, 8 lie between 20 and 25 g and 6 between 37 and 40 g. The first column covering 20–25 g therefore has a height of 1.6. The last column covers the range 37–40 g and has a height of 2. All other intervals are 1 g (see Figure 1.3).

Exercises

1. In a litter of 10 young the probability that any number (in the range 0–10) will be male is given by the following table.

No. of males	0	1	2	3	4	5
Probability	0.001	0.010	0.044	0.117	0.205	0.246
No. of males	6	7	8	9	10	
Probability	0.205	0.117	0.044	0.010	0.001	

Draw a histogram of this probability distribution.

2. During some trials of an insecticide insects of a given species are subjected to increasing concentrations of the insecticide. The increase in mortality is noted for each increase in concentration.

Concentration (ppm)	0–5	5–6	6–7	7–8	8–9	9–10
Mortality (per cent)	3.5	6.0	9.0	10.5	9.5	9.0
Concentration (ppm)	10–15	15–20	20–25	25–30	30–50	
Mortality (per cent)	29.0	12.5	6.0	3.0	2.0	

Draw a histogram to illustrate the mortality distribution with increase in concentration.

3. Draw a histogram to illustrate your hours of study, including lectures, tutorials, and practicals, for each day of the week.

1.2. Graphs

Graphs are used to illustrate data which, for the most part, are changing continuously with respect to some variable. A continuous change can be regarded as a change without sudden jumps. There is a smooth transition from one value to the next. Examples are, continuous recordings of barometric pressure (barograph), continuous temperature recordings (thermograph), variation of reaction rates with concentrations, and variation of electrical potentials down the length of nerve fibres. In the first two examples the variable mentioned above is time, in the third it is concentration, and in the fourth it is distance down the nerve fibres.

In addition, data which may not be strictly continuous can be represented graphically. The growth of a population with time can only take place in whole numbers. However if the population is large enough these individual increases may be occurring very frequently, each one representing a very small fraction of the population as a whole. In this situation the discrete increases merge to such an extent that a continuous representation can be used.

Data which may be expressed in the form of histograms starts to look continuous if it is recorded over shorter and shorter intervals. This can be illustrated by considering the virus infection example of the previous section. If the population were very much larger and infections were reported every hour there would be 480 blocks in the diagram and the profile produced by them would look smoother. It would still have the same general features as the previous histograms. A still shorter interval would yield a profile indistinguishable from a continuous curve (Figure 1.4). One reason for trying to express data in continuous form is that it is more amenable to mathematical analysis.

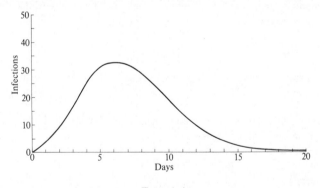

FIG. 1.4

1.3. Rectangular Cartesian coordinates

In order to represent data graphically a coordinate system and associated scales must be defined. The *rectangular Cartesian coordinate system* is the most common. This consists of a pair of perpendicular scaled lines or *axes*, the scaling depending upon the data to be represented.

The intersection of the axes is known as the *origin*. It is conventional to draw one axis horizontal and the other vertical. This divides the graph into *four quadrants* which are numbered as illustrated in Figure 1.5.

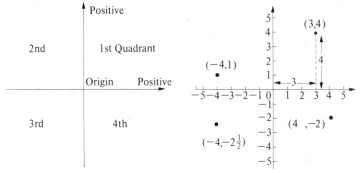

FIG. 1.5

The most commonly used quadrant is the top right and so this is labelled as the *first quadrant*. In mathematics an *anti-clockwise* rotation is *positive* and so the numbering sequence follows.

The axes are scaled with zero at the origin, the positive direction on the horizontal axis being to the right and on the vertical axis being upwards. The position of any point can then be specified by stating its horizontal and vertical displacements from the origin. Conventionally the horizontal displacement or *abscissa* is stated first followed by the vertical displacement or *ordinate*. This gives rise to an *ordered pair* of numbers in which not only the numbers have a meaning but the order in which they are stated is also of significance.

1.4. Independent and dependent variables

When an experiment is being conducted or observations made the person (or machine) recording the results or values usually does so at predetermined times. In this respect time may be regarded as under the control of the observer. In a different situation an insecticide may be under test and various concentrations are administered to aphids, for example, and the percentage mortality noted. Here the concentration administered is under the control of the experimenter. In both cases there is a variable (time or concentration) which in some sense is under the control of the observer. In most experimental situations this is so and it is conventional to associate the variable with the horizontal axis.

This is known as the *independent variable*. The observed variable is known as the *dependent variable* and is associated with the vertical axis. This convention standardizes the way in which graphs are drawn. Once this has been done terms such as 'increasing', 'decreasing', 'reaches a maximum', etc. have fairly well-defined meanings when used to describe graphs. Such descriptions of graphs can help to establish qualitative relationships between variables which may not be apparent from tables of results alone.

Example. The number of organisms in a population, measured in millions, changes with time according to the following data.

Time (h)	0	5	10	15	20	25
Population (millions)	0.84	1.20	1.98	2.88	3.95	5.11

Time (h)	30	35	40	45	50
Population (millions)	6.20	7.15	7.87	8.38	8.72

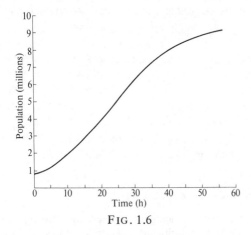

FIG. 1.6

Draw a graph of the population against time and describe how it depends upon time (see Figure 1.6).

Initially the *rate* of growth is rather small but increases and reaches a maximum after about 22 hours. At this point the graph is steepest. It then decreases and the population appears to be levelling out at about 10 million, a ten-fold increase on its original value. Since the population is large a continuous variation has been assumed.

Exercises

1. In the following pairs of recordings which should be regarded as the independent variable and why?
(a) Variation of pH of an alkaline solution as acid is run into it.
(b) Recordings of pulse rate at various intervals after exercise.
(c) Changes in respiration rate for various oxygen proportions in an oxygen/nitrogen atmosphere.
(d) Reaction rates and associated temperatures.
(e) Temperatures and associated reaction rates.

2. A driver's blood-pressure is monitored as he approaches, drives round, and leaves a roundabout. The following recordings are made.

Time (s)	0	1	2	3	4	5	6	7	8	9	10
Pressure (mm Hg)	92	94	99	109	115	118	119	110	102	99	97

Draw the graph of pressure against time and describe the variation of blood-pressure as the driver passes round the roundabout.

3. A solution in one compartment is separated from pure solvent in a second compartment by a membrane. The second is separated from a third, which also contains pure solvent, by another membrane. Solute from the first compartment diffuses across the first membrane into the second and then across the second membrane into the third. Recordings of the concentration of the solution in each of the three compartments were made over a period of several hours. The following results were obtained.

Time (h)	0	1	2	3	4	5	6	7
Concentration (g/l)								
Compartment 1	60	48	39	34	29	27	25	24
Compartment 2	0	17	24	27	26	24	23	22
Compartment 3	0	2	5	10	14	16	17	18

On the same graph plot variations of concentration with time for each of the three compartments. Describe the graphs with particular reference to their dependence on each other.

1.5. Equations

Graphs can be of great value in illustrating the general form of a relationship between variables but they have their limitations. It is difficult to represent accurately on a graph wide variations in the magnitudes of the variables. In addition graphs are specific to the range of values displayed and may give no hint at all about what happens outside the range. If the relationship between a pair of variables is sufficiently well understood a more useful way of expressing it is through an *equation*, the usual form of which is

Dependent variable = Some mathematical expression containing the Independent variable.

For example, the area A of a circle is related to its radius r by

$$A = \pi r^2$$

where $\pi = 3.14159. \ldots$

An equation is a generalization from which predictions can be made about ranges of variable for which no experimental data exist. The form it takes may give some indication of the underlying physical situation it represents. It is amenable to mathematical manipulation which can yield such information as rates of increase or decrease, the positions and values of any maxima or minima, mean values, etc. For such reasons considerable effort in any science is devoted to searching for equations which accurately represent observed phenomena. In the form above the dependent variable can immediately be calculated once the value of the independent variable has been specified.

Example. The mass of a radioactive isotope with a half-life T at any time t is given by the equation

$$m = m_0(\tfrac{1}{2})^{\frac{t}{T}}$$

where m_0 is the mass when $t = 0$.

If $m_0 = 0.1$ g and $T = 41$ min, find the mass remaining after one hour.

From the equation the required mass is

$$m = 0.1 \times (\tfrac{1}{2})^{\frac{60}{41}} = 0.03626 \text{ g.}$$

In view of the fact that the original mass is only given to one significant figure $m = 0.04$ g is a more reasonable way to quote the answer.

Equations can be formulated in a number of ways. If they are formulated solely from experimental data they are said to be *empirical* and give rise to *empirical laws*. Once such laws have been established they often lead to an understanding of the basic principles behind them. If equations arise from untested theory they are said to be *hypothetical* and give rise to *hypotheses*. They are then used to predict the results of experiments and if verified they support the theory which led to their formulation. In the first case an equation is sought with a graph of the same shape as that of some experiments and in the second case an equation is used to predict the shape of the graph of some future experiments. In either case an intimate knowledge of equations and their graphs is invaluable.

In order to obtain this knowledge of equations and their graphs it is useful to be able to generalize away from particular examples. To avoid stating what the independent and dependent variables refer to, use is made of letters to represent them which have no obvious physical interpretation. In mathematics the independent variable is conventionally taken to be x and the dependent variable to be y. Letters near the end of the alphabet are normally regarded as variables and those near the beginning as constants. An equation of the radioactive-decay-type might then be written

$$y = ab^{cx}.$$

The generalization in this case is particularly useful since population growth, cooling under forced convection, the decay of electric charge on a capacitor (nerve cell), and the decrease in light intensity as it passes through a solution all have the same generalized equation. All of these phenomena can therefore be investigated by considering a single equation and its properties.

Example. The volume of a sphere of radius r is given by $V = \frac{4}{3}\pi r^3$ and the volume of a cube of side s is given by $V = s^3$. Are these two formulae of the same general type?

Yes, both generalize to $y = ax^3$. In the first case the constant is $\frac{4}{3}\pi$ and in the second case it is 1.

Example. The volume of blood of viscosity η flowing along a vein of internal radius r and length L in unit time is given by

$$V = \frac{1}{8}\pi\frac{Pr^4}{\eta L}$$

where P is the pressure difference across the ends. If the blood is passed down veins of equal length but different radii with the same pressure difference across the ends, what is the general form of the relationship?

The relationship now only depends upon radius r since all other factors are held constant. Therefore,

$$V = \left(\frac{\pi P}{8\eta L}\right) r^4$$

where $\left(\dfrac{\pi P}{8\eta L}\right)$ is constant. The general form of the relationship is therefore

$$y = ax^4.$$

Example. If in the previous example the radius is held constant and only the length varied what is the general form of the relationship?

The relationship can now be written

$$V = \left(\frac{\pi P r^4}{8\eta}\right) \frac{1}{L}$$

where $\left(\dfrac{\pi P r^4}{8\eta}\right)$ is constant. The general form is therefore

$$y = \frac{b}{x} \text{ or } bx^{-1}.$$

Exercises

1. The pH of a solution is given by the equation

$$pH = -\log[H^+]$$

where $[H^+]$ is the concentration of hydrogen ions in gram moles per litre.
  Find (a) the pH of a solution in which $[H^+] = 0.000275$; (b) $[H^+]$ if the pH is 9.23.

2. The growth of a population in its early stages can be represented by the equation

$$x = x_0 e^{kt}$$

where x_0 is the number at time $t = 0$, k is a constant, and e is the number 2.71828......, a never-ending decimal of the same type as π. If the initial population is 1.35 million, $k = 0.025$, and t is in hours find the size of the population after 100 hours.

3. The radioactive decay of an isotope of half-life T and initial mass m_0 may be represented by

$$m = m_0(\tfrac{1}{2})^{\frac{t}{T}}$$

where t is time measured in the same units as T. Show that if the constant k is chosen correctly the above equation is the same as

$$m = m_0 e^{-kt}$$

where e = 2.71828......

4. The radius r of a circle of area A is given by

$$r = \sqrt{\left(\frac{A}{\pi}\right)}.$$

The periodic time T of a simple pendulum of length L is

$$T = 2\pi \sqrt{\left(\frac{L}{g}\right)}$$

where g is the gravitational acceleration. What is the general form of both of these relationships?

5. A small sphere of radius r and density ρ falling through a viscous medium of density ρ_0 and viscosity η reaches a terminal speed v given by

$$v = \frac{2}{9}gr^2\left(\frac{\rho - \rho_0}{\eta}\right),$$

where g is the gravitational acceleration. What are the general forms that this equation takes if each of the variables r, ρ, ρ_0, and η are regarded in turn as the independent variable? In each case consider all other variables on the right-hand side to be constant.

2 Common functions

2.1. Functions

THE WORD *function* has a particular significance in mathematics which should be understood before this chapter can be appreciated. Such remarks as 'The British character is a function of the British climate', 'a child's intelligence is a function of its environment', or 'a nation's prosperity is a function of its investment' all convey the mathematician's use of the word. Here function means 'influenced by' or 'depends upon' and leads to the idea of the dependent variable encountered in §1.4. A mathematician might say that 'y is a function of x'. What is meant here is that a change in x is known to produce a change in y, even if the precise change cannot be predicted. If the change can be predicted then some functional relationship between x and y exists which is often expressed in the form of an equation.

It is the purpose of subsequent sections in this chapter to look at the more common functional relationships and their graphs and how (if at all) they are related to one another in order to develop in the reader an ability to 'picture' the shape of a function if its mathematical form is given and, conversely, to make a reasonable guess at the mathematical form if a graph is given.

2.2. Linear functions

The simplest relationship between x and y is having y take on the same value as x. The equation describing this is

$$y = x$$

and a table of values of y for x values between -2 and $+2$ at unit intervals is

x	-2	-1	0	1	2
y	-2	-1	0	1	2

The corresponding ordered pairs (cf. §1.3) are $(-2, -2)$, $(-1, -1)$, $(0, 0)$, $(1, 1)$, $(2, 2)$ and can be plotted on a pair of rectangular Cartesian coordinate axes. If this is done it will be noticed that all five points lie on a straight line which passes through the origin and slopes upwards from left to right. At this stage there is a natural tendency to take a ruler and draw a straight line through all five points and, without thinking, to regard this as the graph of the equation $y = x$ (Figure 2.1(a)). In fact an assumption has been made. The assumption is that the simplest line or curve passing through a set of points is the correct one unless anything is known to the contrary. This is just a particular

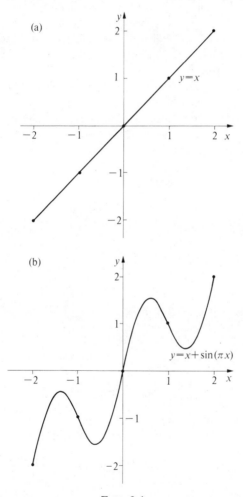

FIG. 2.1

case of the general principle that the simplest explanation of a phenomenon which satisfies all the facts is the one to be adopted. For the
equation $y = x$ it is easy to check that intermediate points do indeed
lie on the line as do points beyond each of the end-points $(-2, -2)$
and $(2, 2)$. However, it is perhaps worth mentioning that the equation

$$y = x + \sin(\pi x),$$

produced by adding to the right-hand side the trigonometric function
$\sin(\pi x)$, has the same table of values as above but its graph (Figure
2.1(b)) is not a straight line. Trigonometric functions are considered in
§2.8. Care should therefore be taken in drawing or sketching graphs.
Although the general principle mentioned above is sound it may lead
to errors if insufficient data or over-selective data are used. In the
examples above the fact that only integers (whole numbers) were used
for x led to identical tables since $\sin(\pi x)$ happens to be zero for all
integer values of x. Had some fractions been used, a difference between
the two tables of values would have been apparent. Intuition might
have led the reader to expect the more complex of the two equations
to have a more complex graph.

A similar table of values for the equation

$$y = 2x + 1$$

is

x	-2	-1	0	1	2
y	-3	-1	1	3	5

and the graph is shown in Figure 2.2. The values of y increase more
quickly from left to right than for the line $y = x$ and $y = 2x + 1$ does
not pass through the origin.

Both $y = x$ and $y = 2x + 1$ are examples of the more general relationship

$$y = mx + c$$

where m and c are constants. This is known as a *linear relationship*
and the function $mx + c$ as a *linear function* since all such equations
describe *straight lines*. It is quite simple to show that a unit increase
in the independent variable x produces an increase of m units in the
dependent variable y. Assume that the initial value of the independent
variable is x_0. The corresponding value of the dependent variable is then

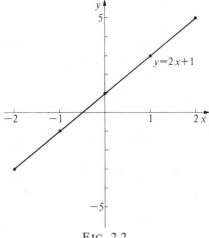

$$y = 2x + 1$$

FIG. 2.2

$$y_0 = mx_0 + c.$$

If the independent variable is now increased by one unit to $x_0 + 1$ the dependent variable becomes

$$y = m(x_0 + 1) + c.$$

The increase in the dependent variable is therefore

$$y - y_0 = \{m(x_0 + 1) + c\} - \{mx_0 + c\} = m.$$

If m is negative, there is a *negative increase* or *decrease* in the dependent variable. The constant m is in fact the *slope* or *gradient* of the line which can be defined to be the *increase* in the dependent variable (y) resulting from a *unit increase* in the independent variable (x).

Similarly it is also quite simple to ascertain the significance of the constant c. At all points on the y-axis, $x = 0$. Putting $x = 0$ into $y = mx + c$ therefore gives the value of y at which the line $y = mx + c$ intercepts the y-axis. Clearly this is at $y = c$. The constant c is therefore the y-axis *intercept*.

There are two sets of parallel lines which do not take the form $y = mx + c$. If the line $y = mx + c$ is considered where m is very small then the graph is of a nearly horizontal line passing through the y-axis at $(0, c)$. If m tends to zero the line becomes horizontal and its equation is

$$y = c.$$

In particular if $c = 0$ it is a horizontal line passing through the origin. It is therefore the x-axis and has equation $y = 0$. Similarly the set of equations

$$x = c$$

for different values of c represent a set of vertical lines which cross the x-axis at $(c, 0)$. In particular the line with $c = 0$ is the y-axis with equation $x = 0$.

If the slope m and vertical axis intercept c are known for a line, its equation can be written as

$$y = mx + c.$$

In addition such a line can be drawn using just m and c since one point $(0, c)$ is known and the slope of the line is known. A ruler can be placed with its edge through $(0, c)$ and rotated until it has slope m and the line drawn. More generally if any single point on a line is known and its slope is also known it can be drawn with a ruler in the same way. Alternatively if two points are known on the line a ruler can be placed with its edge through both and the line drawn. Since each of these methods of drawing a line is possible it should also be possible to find the equation of a line given either a single point on it and its slope or two points on it. Furthermore these two pairs of starting conditions ought to be equivalent to each other since a given line can be drawn by either method.

To pursue this further we will consider a specific example. Assume we are required to find the equation of the line which passes through $(2, 4)$ and has slope 3. A sketch of the situation can be helpful and is shown in Figure 2.3. It is clear from this that the y-axis intercept is below the origin and is therefore negative. It is equally clear that the line crosses the x-axis between $x = 0$ and $x = 1$. Since the slope is given, we know $m = 3$ so the equation has the form

$$y = 3x + c$$

where c is unknown. We also know that the required line passes through $(2, 4)$ and substitution of these coordinates into the equation leaves c as the only unknown. Therefore,

$$4 = 3 \times 2 + c$$

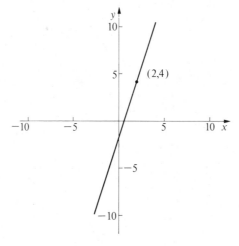

FIG. 2.3

and so

$$c = -2.$$

The equation is therefore

$$y = 3x - 2.$$

The original information about one point and the slope can be converted into information about two points quite simply. The given point is $(2, 4)$ and the slope is 3. Hence a unit increase in x produces a three-unit increase in y. A second point is therefore $(3, 7)$ and the line can equally well be specified as that line which passes through $(2, 4)$ and $(3, 7)$ rather than as the line passing through $(2, 4)$ with slope 3.

If the starting information were two specified points, the procedure is slightly longer. Assume that we require the equation of the straight line passing through $(3, 1)$ and $(9, -1)$. Again a sketch might prove useful and is shown in Figure 2.4. The line has negative slope, crosses the x-axis at or near $x = 6$ and the y-axis at or near $y = 2$. We assume the equation is

$$y = mx + c$$

where we are required to find m and c. Since the line is to pass through each of $(3, 1)$ and $(9, -1)$ then both of the following equations must be satisfied simultaneously.

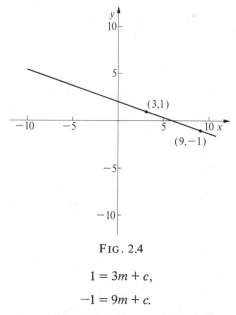

F$_\text{IG}$. 2.4

$$1 = 3m + c,$$

$$-1 = 9m + c.$$

Subtracting the first from the second eliminates c and shows that $m = -\frac{1}{3}$. The slope is now known and either of the known points can be used to find c. Hence, either,

$$1 = -\tfrac{1}{3} \times 3 + c \quad \text{and so} \quad c = 2,$$

or

$$-1 = -\tfrac{1}{3} \times 9 + c \quad \text{and so} \quad c = 2.$$

The equation is therefore

$$y = -\tfrac{1}{3}x + 2.$$

The slope m could have been found directly from the two points. As x increases from 3 to 9, y decreases from 1 to -1. In other words for a 6-unit increase in x there is a -2-unit increase in y. Hence a unit increase in x produces $-\frac{1}{3}$ of a unit increase in y. The slope is therefore $-\frac{1}{3}$. Either point can now be used to determine c and the situation is one of a knowledge of a single point and the slope.

The equation may also be obtained directly from a knowledge of two points by reference to a third but arbitrary point, (x, y), which is to lie on the line. The slope can then be calculated from the two known points and also from a single known point and the arbitrary point. If all three points lie on the line then these two slopes must be the same and the equation is formed. If the two points $(3, 1)$ and $(9, -1)$ are used together with the arbitrary point (x, y) as shown in Figure 2.5, the slope using the two known points is given by

$$\frac{(-1) - 1}{9 - 3} = -\tfrac{1}{3}.$$

Using the first point $(3, 1)$ and the arbitrary point (x, y) the slope is recalculated to be

$$\frac{y - 1}{x - 3}.$$

These two slopes must be the same and so

$$\frac{y - 1}{x - 3} = -\tfrac{1}{3},$$

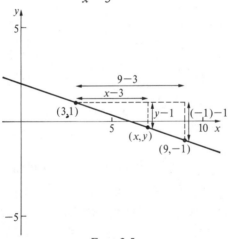

FIG. 2.5

$$y - 1 = -\tfrac{1}{3}(x - 3),$$
$$y = -\tfrac{1}{3}x + 2.$$

If the two known points are (x_1, y_1) and (x_2, y_2) and an arbitrary third point (x, y) is included then reference to Figure 2.6 shows that the two equal slopes are

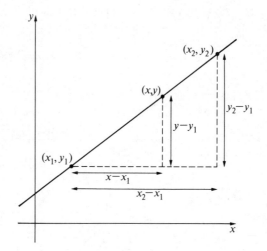

F IG. 2.6

$$\frac{y_2 - y_1}{x_2 - x_1} \quad \text{and} \quad \frac{y - y_1}{x - x_1}.$$

The equation is therefore

$$\frac{y - y_1}{x - x_1} = \frac{y_2 - y_1}{x_2 - x_1}$$

or

$$y - y_1 = \left(\frac{y_2 - y_1}{x_2 - x_1}\right)(x - x_1).$$

If a single point, (x_1, y_1), and the slope m are known then the method is slightly shorter since the slopes are now

$$m \quad \text{and} \quad \frac{y - y_1}{x - x_1}.$$

The equation in this case is

$$y - y_1 = m(x - x_1).$$

Example. The percentage germination of seeds depends upon temperature. It is found that for a variety of tomato seeds 40% germinate at 12°C and 70% germinate at 15°C. If the relationship between percentage germination G and temperature T is assumed to be linear find G as a function of T. Hence determine the germination at 10°C and the temperature required for 90% germination.

Using the formula above the equation of the line passing through two known points can be written down. If the germination G is the dependent variable we have

$$G - 40 = \left(\frac{70 - 40}{15 - 12}\right)(T - 12)$$

and so

$$G = 10T - 80 = 10(T - 8).$$

The germination at 10°C is therefore 20% and rewriting the equation in the form

$$T = \frac{1}{10}(G + 80)$$

it can be seen that the temperature required for 90% germination is 17°C.

Exercises

1. Sketch the lines given by the equations

(a) $y = 3x$;   (b) $y = 3x + 1$;   (c) $y = 1 - 2x$;

(d) $y = 2x - 3$;   (e) $y = 3 - 2x$;   (f) $y = 3$;

(g) $y = -2$;   (h) $x = 2$;   (i) $x = -5$;

(j) $x = 2 - 3y$.

2. Find the equations of the lines which pass through the given point and have the given slope

(a) $(1, 1)$ slope 1;   (b) $(0, 2)$ slope $\frac{1}{2}$;   (c) $(-1, 2)$ slope -2;

(d) $(5, 2)$ slope $-\frac{1}{3}$;   (e) $(4, 2)$ slope 0.

3. Find the equations of the lines which pass through the following pairs of points.

(a) $(1, 2)$ and $(3, 4)$;   (b) $(0, 0)$ and $(1, 2)$;

(c) $(-1, 5)$ and $(2, 2)$;   (d) $(a, 0)$ and $(0, b)$;

(e) (a, b) and (b, a);   (f) $(2, 4)$ and $(5, 4)$;

(g) $(-1, 5)$ and $(3, 5)$;   (h) $(3, 1)$ and $(3, 7)$.

4. Using similar triangles or otherwise prove that the perpendicular to a line of slope m has slope $-1/m$.

5. Find the equation of the line through $(3, 1)$ which is perpendicular to the line $y = x + 2$. Find the point of intersection of these lines and hence the perpendicular distance from $(3, 1)$ to the line $y = x + 2$.

6. Find the perpendicular distance from $(-2, 3)$ to the line $y = 2x + 3$.

2.3. Quadratic functions

Having discussed linear functions in the previous section we now consider functions in which the highest power of x is x^2. Such a function is known as a *quadratic function* of x and has the general form

$$ax^2 + bx + c,$$

where a, b, and c are constants. For the function to be a quadratic the constant a cannot be zero; otherwise there is no x^2 and only a linear function remains.

The simplest quadratic is x^2 ($a = 1, b = 0, c = 0$) and the associated equation is

$$y = x^2.$$

Note that in the equation if x is replaced by $-x$ the equation is unchanged. This means that the same value of y occurs for values of x which are equal in magnitude or size but opposite in sign. This is clear from the following table of values for x between -3 and $+3$ in unit steps.

x	-3	-2	-1	0	1	2	3
y	9	4	1	0	1	4	9

Since values of x which are equal in magnitude but opposite in sign lie at equal distances from the y-axis but on opposite sides of it and each gives the same value of y, the graph of $y = x^2$ is symmetrical about the y-axis as shown in Figure 2.7. Such functions are known as *even functions* of x. Furthermore, y can never be negative in this case

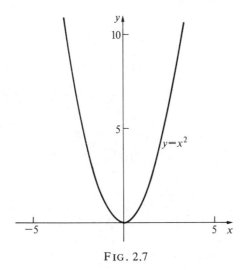

$$y = x^2$$

FIG. 2.7

because the square of both positive and negative values of x is still positive and only if $x = 0$ can $y = 0$. The graph therefore lies wholly above the x-axis except for the single point at the origin. The shape of the graph, which is characteristic of quadratic functions, is known as a *parabola*.

Drawing the graphs of a number of different quadratic functions illustrates their common shape and how different combinations of the constants a, b, and c affect the positions of the graphs. The graphs of the equations,

$$y = x^2, y = 2x^2, y = x^2 - 3x + 2,$$

$$y = x^2 + 2x + 2, y = 4 - x^2, y = -x^2 - 2x - 1,$$

are shown in Figure 2.8. It is worth drawing the reader's attention to a number of points about these graphs. All are the same shape but if the coefficient of x^2 is negative, as in the last two examples, the parabola is 'nose up', rather than the more conventional 'nose down'. If the coefficient of x is zero, that is if there is no term in x, then the parabola is symmetrical about the y-axis. Putting $x = 0$ gives the point at which the parabola crosses the y-axis. Putting $y = 0$ gives a quadratic equation which will either have two different solutions, two identical solutions, or no solutions at all in terms of conventional numbers. In the first case, the parabola crosses the x-axis at two points, in the second it

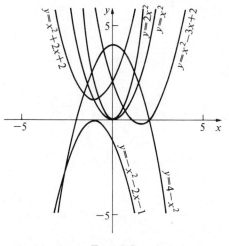

F IG. 2.8

just touches it at a single point, and in the third it misses it altogther.

By putting $y = 0$ in any one of the above equations we are asking 'Where does the line $y = 0$ (the x-axis) cross the parabola?' To find the answer we are solving simultaneously the equation $y = 0$ and the equation of the parabola. As an example consider the pair

$$y = 0,$$
$$y = x^2 - 3x + 2.$$

Substituting the first into the second gives the quadratic equation

$$x^2 - 3x + 2 = 0.$$

This can be solved by factorization or by use of 'the formula' for quadratic equations. The factors are $(x - 1)$ and $(x - 2)$ so that

$$(x - 1)(x - 2) = 0$$

and $x = 1$ or 2. The parabola $y = x^2 - 3x + 2$ therefore crosses the x-axis ($y = 0$) at $x = 1$ and $x = 2$ (Figure 2.8).

Considering the last of the equations we wish to solve simultaneously

$$y = 0 \quad \text{and} \quad y = -x^2 - 2x - 1.$$

This gives the quadratic equation $x^2 + 2x + 1 = 0$ which factorizes to give $(x + 1)^2 = 0$. There are therefore two identical solutions of $x = -1$. The conclusion is that the parabola $y = -x^2 - 2x - 1$ just touches the x-axis at $x = -1$ (Figure 2.8).

The third possible outcome is exemplified by the parabola $y = x^2 + 2x + 2$. Solving simultaneously with $y = 0$ gives the quadratic equation

$$x^2 + 2x + 2 = 0.$$

There are no obvious factors here and use of 'the formula' shows that

$$x = \frac{-2 \pm \sqrt{(4 - 4 \times 1 \times 2)}}{2}$$

$$= -1 \pm \sqrt{-1}.$$

The $\sqrt{-1}$ cannot be interpreted in terms of conventional numbers and so we conclude that the parabola does not cross or touch the x-axis at all. The graph in Figure 2.8 confirms this to be correct. Note however that even though the quadratic equation has no conventional solution, the solution obtained does hold some useful information. If we neglect the non-conventional part, that is $\sqrt{-1}$, the remaining part, -1, gives the vertical line, $x = -1$, about which the parabola $y = x^2 + 2x + 2$ is symmetrical.

More generally if the equation

$$y = ax^2 + bx + c$$

is considered and solved simultaneously with $y = 0$ (the x-axis) the quadratic equation

$$ax^2 + bx + c = 0$$

is obtained which has the general solution

$$x = \frac{-b \pm \sqrt{(b^2 - 4ac)}}{2a}$$

This is 'the formula used so frequently to solve quadratic equations. A proof of this result is given in Appendix A.3. The expression

$$b^2 - 4ac$$

is known as the *discriminant*. Whether it is positive, zero, or negative determines the type of solution and hence the type of intersection, if any, of the parabola with the x-axis. If $b^2 - 4ac > 0$ then its square root can be found and 'the formula' gives two distinct solutions which are symmetrical about $x = -b/2a$. The parabola therefore crosses and then recrosses the x-axis. If $b^2 - 4ac = 0$ then there is only the single solution $x = -b/2a$ and the parabola just touches the x-axis. The line of symmetry of the parabola also passes through this point. If $b^2 - 4ac < 0$, there is no conventional square root since the number inside it is negative and so the parabola does not cross or touch the x-axis. It is, however, symmetrical about $x = -b/2a$. Since the line of symmetry of a parabola passes through its lowest point if it is 'nose down' and its highest point if it is 'nose up', putting $x = -b/2a$ into $y = ax^2 + bx + c$ gives the position of such points. The coordinates are $(-b/2a, c - (b^2/4a))$.

Example. Sketch the graph of $y = x^2 - \tfrac{5}{2}x + 1$.

The right-hand side is a quadratic function and so the graph is a parabola. Putting $x = 0$ shows that it crosses the y-axis at $y = 1$ and putting $y = 0$ leads to the equation $x^2 - \tfrac{5}{2}x + 1 = 0$. Using 'the formula' gives

$$x = \frac{\tfrac{5}{2} \pm \sqrt{\{(\tfrac{5}{2})^2 - 4\}}}{2} = \tfrac{1}{4}(5 \pm 3).$$

Hence the line of symmetry is $x = \tfrac{5}{4}$ and the parabola crosses the x-axis at $x = \tfrac{1}{2}$ and $x = 2$. The graph is shown in Figure 2.9.

Example. Sketch the graph of $y = -x^2 + 4x - 5$.

The graph is a parabola and, since the coefficient of x^2 is negative, it is 'nose up'. It crosses the y-axis at $y = -5$. Putting $y = 0$ gives $x^2 - 4x + 5 = 0$ which has solutions

$$x = \frac{4 \pm \sqrt{(16 - 20)}}{2} = 2 \pm \sqrt{-1}.$$

The parabola does not cross the x-axis and is symmetrical about $x = 2$. Its highest point is $(2, -1)$ and the graph is shown in Figure 2.10.

The general quadratic function has associated with it the equation

$$y = ax^2 + bx + c$$

containing three constants a, b, and c. In the same way that two pieces

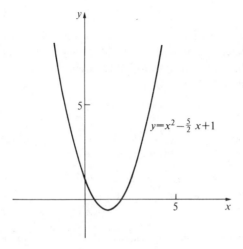

$$y = x^2 - \tfrac{5}{2}x + 1$$

FIG. 2.9

of information were required to find m and c in the general linear equation, $y = mx + c$, we require three pieces of information for the general quadratic to determine a, b, and c. This information can be knowledge of three points through which the parabola is to pass.

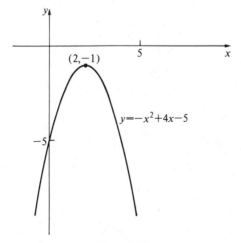

$(2,-1)$

$$y = -x^2 + 4x - 5$$

FIG. 2.10

Example. Find the equation of the parabola which passes through $(-1, 0), (1, 6)$, and $(5, -6)$.

The general equation $y = ax^2 + bx + c$ must be satisfied at each of the three points simultaneously. The following three equations, formed by substituting each point in turn into the general equation, must therefore be solved simultaneously.

$$0 = a - b + c,$$
$$6 = a + b + c,$$
$$-6 = 25a + 5b + c.$$

Subtracting the first from the second gives $b = 3$. Subtracting the second from the third gives

$$-12 = 24a + 4b$$

and hence $a = -1$ since $b = 3$. Finally $c = 4$ and the equation is

$$y = 4 + 3x - x^2.$$

Its graph is shown in Figure 2.11.

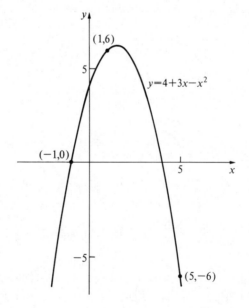

FIG. 2.11

Example. The percentage germination G of a variety of tomato is measured at three temperatures T. At $9\,^{\circ}$C only 20% germinate, at $12\,^{\circ}$C the figure is 40% and at $15\,^{\circ}$C it is 70%. Find an expression for G as a quadratic function of T and using this expression predict and comment upon the percentage germination at $17\,^{\circ}$C and $18\,^{\circ}$C. At what temperature is the germination 10%?

If G is to be a quadratic function of temperature T then

$$G = aT^2 + bT + c.$$

This quadratic is to pass through the three points (9, 20), (12, 40), and (15, 70) and so the following three equations must be solved simultaneously.

$$20 = 81a + 9b + c,$$
$$40 = 144a + 12b + c,$$
$$70 = 225a + 15b + c.$$

Subtracting the second from the third and the first from the second gives the two equations

$$30 = 81a + 3b,$$
$$20 = 63a + 3b,$$

from which c has been eliminated. Subtraction of the second of these from the first eliminates b and gives

$$10 = 18a.$$

Hence $a = \frac{5}{9}$ and so $b = -5$ and $c = 20$. The relationship is therefore

$$G = \tfrac{5}{9}T^2 - 5T + 20 = \tfrac{5}{9}(T^2 - 9T + 36).$$

When $T = 17$ the germination G is 96% and when $T = 18$ it is 110%. This last figure is clearly impossible and so the equation must be used with care. It may well have rather restricted validity.

To find at what temperature the germination is 10% we must solve the equation

$$10 = \tfrac{5}{9}T^2 - 5T + 20$$

or

$$T^2 - 9T + 18 = 0.$$

The solutions are given by (Appendix A.3)

$$T = \frac{9 \pm \sqrt{(81 - 4 \times 1 \times 18)}}{2}$$

$$= \tfrac{1}{2}(9 \pm 3) = 6 \text{ or } 3.$$

The figure of $3\,^{\circ}\text{C}$ is unrealistic. The value of $6\,^{\circ}\text{C}$ fits in better with the original measured values of temperature and germination. The reason for two values emerging from the calculation can be seen if a graph is drawn (Figure 2.12). It has a minimum of about 9% germination at a temperature of $4.5\,^{\circ}\text{C}$. Indeed, at $0\,^{\circ}\text{C}$ the germination has risen to 20% according to the formula. This unrealistic value again points to the restricted validity of the expression.

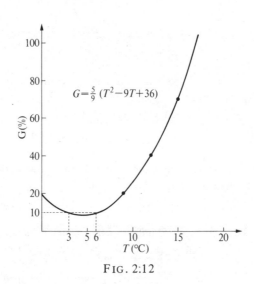

$$G = \tfrac{5}{9}\,(T^2 - 9T + 36)$$

FIG. 2:12

Exercises

1. Sketch the graphs of the following equations.

(a) $y = x^2 - 4$;   (b) $y = 4 - x^2$;   (c) $y = x^2 - 2x - 3$;

(d) $y = 2 - x - x^2$;   (e) $y = 2x^2 - x - 1$;   (f) $y = x^2 - 2x + 1$;

(g) $y = 4x^2 - 12x + 9$;   (h) $y = x^2 - 2x + 2$;

(i) $y = -x^2 + 2x - 3$;   (j) $y = 4 - x - 3x^2$.

2. Find the equations of the parabolas which pass through the following sets of points and hence sketch their graphs.

(a) $(2, -3), (3, 0), (4, 5)$;

(b) $(-3, 5), (-1, 1), (4, 26)$;

(c) $(-2, -3), (-1, 2), (2, -7)$;

(d) $(0.50, -1.75), (1.60, 0.56), (3.50, 10.25)$;

(e) $(1.00, 3.70), (2.50, -9.50), (3.50, -26.30)$.

3. By making accurate drawings of the graphs of $y = x^2$ and $y = 2x + 3$ find the solutions of the equation $x^2 - 2x - 3 = 0$.

4. By considering the graphs of $y = x^2$ and $y = 10 - 3x$, find the solutions of $x^2 + 3x - 10 = 0$.

5. Two saplings grow at different rates. When recordings are started one is 100 mm high and grows at a rate of 1 per cent of its current height each week. The height after n weeks can be approximated by the expression

$$100 + 0.995n + 0.005n^2.$$

The other sapling is initially 100 mm high and grows at a steady rate of 1 per cent of its initial height each week. Its height after n weeks is given by

$$110 + 1.1n.$$

By drawing graphs or otherwise find to the nearest week the time at which the saplings will have reached the same height. If the same growth expressions are assumed to have held before recordings began, find the time when the saplings were previously at the same height.

6. A length of 800 metres of wire fencing is to be used to form a rectangular enclosure. If the length of one of the sides is assumed to be x metres draw a graph of the area A against x and hence determine the maximum area which can be fenced in. Find also the dimensions of the enclosure in this case. If a straight hedge can be used for one side of the enclosure and the fencing is used for the other three sides draw a graph of A against x corresponding to this new situation. Hence find the new maximum area enclosed and the dimensions required.

7. The rate of growth R of a population of size p which is limited by its environment to an upper bound of size P may be represented by

$$R = kp(P - p)$$

where k is a positive constant. A population of this type is initially observed when its size is half a million. If $k = 2$ and the upper bound is 20 million sketch a graph of R against p. What conclusions can be drawn from this graph about the shape of a graph of population p against time t?

2.4. Polynomial functions

Linear and quadratic functions are the simpler members of a class of functions known as *polynomials*. Included in this class are *cubic functions* which have x^3 as the highest power of x, *quartic functions* with x^4 as the highest power, *quintic functions* (x^5), and so on. A polynomial of *degree n* has x^n as the highest power of x. In its simplest form it may have no other terms at all, in which case it is said to be *power function*, but in general it will contain lower powers of x as well. The general form of a polynomial of degree n is

$$a_n x^n + a_{n-1} x^{n-1} + a_{n-2} x^{n-2} + \ldots \ldots + a_2 x^2 + a_1 x + a_0$$

where all the $a_i, i = 0, 1, \ldots, n$, are coefficients. The only restriction on them is that $a_n \neq 0$ otherwise the term in x^n would vanish and the polynomial would not be of degree n. This general polynomial can be written in a more compact form using the *sigma notation*, Σ. The Greek capital sigma is used to signify addition of terms and so

$$\sum_{i=0}^{n} a_i x^i$$

means add together all terms of the form $a_i x^i$ starting with $i = 0$ and finishing with $i = n$. The expansion is therefore

$$\sum_{i=0}^{n} a_i x^i = a_0 x^0 + a_1 x^1 + a_2 x^2 + a_3 x^3 + \ldots + a_{n-1} x^{n-1} + a_n x^n$$

and since $x^0 = 1$ and $x^1 = x$ we have the same polynomial as above but written in reverse order. The following are some examples of polynomials.

Polynomial	Degree
$x^2 - 3x + 7$	2
$4x^7 - x^4 + 5x^3 + x^2 - 3x + 1$	7
$3x^4 + 7x^3 - 8x^3 + 2x - 3$	4
$2x - 1$	1
x^{10}	10
$x^3 - 3x^2 + 3x + 1$	3

The graphs of linear and quadratic functions have been discussed in §§ 2.2 and 2.3 respectively. The simplest cubic function is x^3 and the associated equation is

$$y = x^3.$$

A typical table of values might be

x	-3	-2	-1	0	1	2	3
y	-27	-8	-1	0	1	8	27

from which the graph shown in Figure 2.13 can be drawn. Some intermediate points near the origin may need to be calculated to show that, as it approaches the origin from the left, it levels off and then

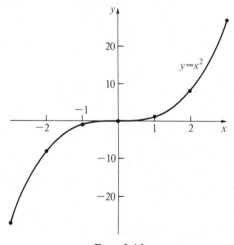

FIG. 2.13

climbs away again to the right. The addition of a constant, for example

$$y = x^3 + 2,$$

simply moves the graph upwards by 2 units. If terms containing x or x^2 or both are included then peaks and troughs may occur. The graphs of

$$y = x^3 + 2,$$

$$y = x^3 - x,$$

$$y = 2x^3 + 3x^2 + 6x,$$

$$y = 1 - 2x - 3x^2 - x^3$$

are shown in Figure 2.14. All are cubics but the shapes of their graphs show considerable variations. For large values of the independent variable, x, the shape is dominated by the term containing x^3. Putting

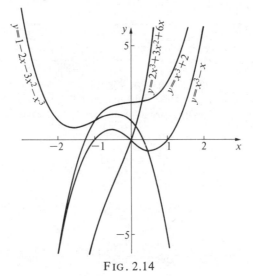

FIG. 2.14

$x = 0$ (the y-axis) gives the point at which each graph crosses the vertical axis. Putting $y = 0$ (the x-axis) gives a cubic equation, the solution of which gives the point or points at which the graph crosses the horizontal axis. The cubic must cross the horizontal axis at least once and may cross it three times. It cannot cross it just twice. The reason for this is that a cubic exhibits a general trend upwards from left to right if the coefficient of x^3 is positive and downwards if the co-efficient is negative. If the trend is upwards it must cross the horizontal axis on the way up. If it does happen to turn downwards at any point it must subsequently turn upwards again to follow the general trend. If this downturn causes it to recross the axis the subsequent upturn will cause it to cross the axis a third time in order to continue the upward trend. A similar argument holds for a cubic with a downward trend.

Example. Sketch the graph of $y = x^3 - 5x^2 + 2x + 8$.

Since the coefficient of x^3 is positive the general trend is upwards from left to right. Putting $x = 0$ shows that it crosses the y-axis at $y = 8$. Putting $y = 0$ gives the equation $x^3 - 5x^2 + 2x + 8 = 0$, the solutions of which give the points (if any) where the graph crosses the x-axis. The left-hand side can be factorized to give $(x + 1)(x - 2)(x - 4) = 0$ and so the points are $x = -1, 2$, and 4. Evaluating the right-hand side of the equation for $x = 1$ and $x = 3$ fills in the intermediate points as $(1, 6)$ and $(3, -4)$. The graph can now be sketched with reasonable accuracy and is shown in Figure 2.15.

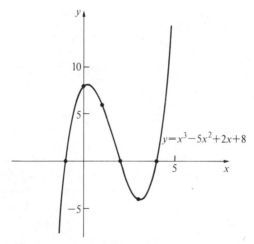

FIG. 2.15

Example. Sketch the graph of $y = 2 + \frac{3}{2}x^2 - \frac{1}{2}x^3$.

The coefficient of x^3 is $-\frac{1}{2}$ and so the trend is downwards from left to right. The graph crosses the y-axis at $(0, 2)$ and the x-axis at the conventional solutions of $2 + \frac{3}{2}x^2 - \frac{1}{2}x^3 = 0$. This equation can be written

$x^3 - 3x^2 - 4 = 0$ which has no obvious factors. The graph will therefore have to be drawn from a table of values with intermediate points calculated in doubtful areas.

x	−3	−2	−1	0	1	2	3	4	5
y	29	12	4	2	3	4	2	−6	−23

The graph is shown in Figure 2.16. More sophisticated methods exist for the location of any peaks and troughs and are discussed in §4.6.

As shown in §2.2 a knowledge of two points on a line is sufficient to find the equation of the line. A knowledge of three points is sufficient

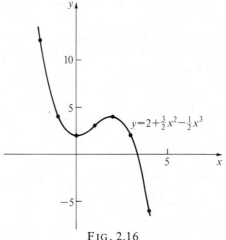

$$y = 2 + \tfrac{3}{2}x^2 - \tfrac{1}{2}x^3$$

FIG. 2.16

to find the equation of a parabola passing through them (cf. §2.3) and it may be expected that four points are required for a cubic equation. Using the notation for the general polynomial a general cubic equation has the form

$$y = a_0 + a_1 x + a_2 x^2 + a_3 x^3.$$

This has four unknown coefficients and so a knowledge of four points gives rise to four equations which can be solved simultaneously for the coefficients $a_0, a_1, a_2,$ and a_3.

Example. A cubic equation is to pass through the points $(-2, -30)$, $(0, -2)$, $(2, 2)$, and $(3, 10)$. Find the equation and sketch its graph.

Using each point in turn we have

$$(-2, -30) \qquad -30 = a_0 - 2a_1 + 4a_2 - 8a_3,$$
$$(0, -2) \qquad -2 = a_0,$$
$$(2, 2) \qquad 2 = a_0 + 2a_1 + 4a_2 + 8a_3,$$
$$(3, 10) \qquad 10 = a_0 + 3a_1 + 9a_2 + 27a_3.$$

Adding the first and third gives $-28 = 2a_0 + 8a_2$ and, since $a_0 = -2$, we have $a_2 = -3$. Substituting these values into the third and fourth equations gives $8 = a_1 + 4a_3$ and $13 = a_1 + 9a_3$. Subtracting the first from the second shows that $a_3 = 1$ and so $a_1 = 4$. The equation is therefore

$$y = -2 + 4x - 3x^2 + x^3.$$

A table of values is

x	-2	-1	0	1	2	3	4
y	-30	-10	-2	0	2	10	30

and from this the graph shown in Figure 2.17 can be sketched.

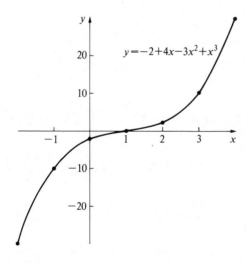

FIG. 2.17

The general quartic has the form

$$a_0 + a_1x + a_2x^2 + a_3x^3 + a_4x^4$$

and is dominated by x^4 when x is large. The graph of $y = x^4$ is shown in Figure 2.18 together with that of a more complex quartic, $y = x^4 - 10x^2 + 9$. Given five points the equation of a quartic which passes through all five can be determined.

The general polynomial of degree n has associated with it the equation

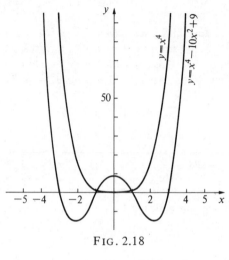

FIG. 2.18

$$y = \sum_{i=0}^{n} a_i x^i$$

which contains $n + 1$ coefficients a_i, $i = 0, 1, \ldots, n$. Its graph crosses the y-axis at $y = a_0$ and when x is large the shape is determined by the term $a_n x^n$. It crosses the x-axis at the conventional solutions of

$$\sum_{i=0}^{n} a_i x^i = 0.$$

Exercises

1. Sketch the graphs of the following equations:

(a) $y = x^3 + 1$;

(b) $y = 1 - x^3$;

(c) $y = (x - 1)(x - 2)(x - 3)$;

(d) $y = x^4 - 2$;

(e) $y = (x^2 - 1)(x^2 - 4)$;

(f) $y = x^5$.

2. Find the equations of the cubics which pass through the following sets of points

(a) $(0, 4), (1, -2), (2, 2), (3, 11)$;

(b) $(-2, -6), (2, 6), (3, 24), (5, 120)$;

(c) $(0.10, -2.56), (0.40, -1.78), (0.90, 4.32), (1.00, 6.80)$.

3. Rectangular open-topped metal containers with square bases are to be fabricated from sheets of metal one metre square. The containers are manufactured by cutting away identical squares of metal from each corner of the original sheet leaving a square central region with a rectangular leaf on each of its sides. The four leaves are then bent upwards to form the sides of the container. Draw a graph of the volume of the container against the base dimension and hence determine the dimensions required to produce the container with maximum volume.

4. Two saplings grow at different rates. When recordings are started one is 100 mm high and grows at a rate of 1 per cent of its current height each week. A good approximation to its height after n weeks is

$$100 + 0.995n + 0.00495n^2 + 0.0000167n^3.$$

The second sapling is initially 110 mm high and grows at a steady rate of 1 per cent of its initial height each week. Its height after n weeks is given by

$$110 + 1.1n.$$

Find the time at which the two saplings reach the same height (cf. §2.3, exercise 6).

5. Recordings of the germination of some seeds at four different temperatures yielded the following results.

Temperature ($^\circ$C)	6	9	12	15
Germination (%)	0	20	40	70

Find an expression for the germination G as a function of temperature T which satisfies the tabulated results.

2.5. Rational functions

A *rational function* is formed when one polynomial is divided by another and, by its very nature, it is more complicated than anything we have met so far. Examples of rational functions are

$$\frac{x-1}{x^2+2}, \quad \frac{1}{x}, \quad \frac{x^4-3x^2+1}{x^3-x-1}, \quad \frac{x}{x^2-4}.$$

Points at which the denominator vanishes are not defined and are excluded since the functions cannot be satisfactorily evaluated at such points. These points are known as *singularities*. The graphs of rational functions can be difficult to sketch because of their behaviour near the singularities.

Example. Sketch the graph of $y = 1/x$.

This is the simplest rational function and has a singularity at $x = 0$. From the equation it can be seen that the sign of y is the same as the sign of x. When x is large, y is small and, when x is small, y is large. A consequence of these properties is that as x passes from negative, through zero (excluded), to positive values y jumps suddenly from being very large negative to being very large positive. A table of values illustrates the point.

x	-5	-2	-1	$-\frac{1}{2}$	$-\frac{1}{5}$	0	$\frac{1}{5}$	$\frac{1}{2}$	1	2	5
y	$-\frac{1}{5}$	$-\frac{1}{2}$	-1	-2	-5	?	5	2	1	$\frac{1}{2}$	$\frac{1}{5}$

The graph is shown in Figure 2.19. It is a *rectangular hyperbola* and like the parabola in §2.3 is one of a group of curves known as *conic*

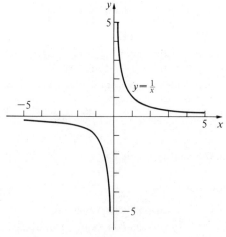

Fig. 2.19

sections. Other members are ellipses and hyperbolae. The circle is a special case of the ellipse in which the major and minor axes are equal and the rectangular hyperbola is a special case of the general hyperbola. The reason for the term 'conic section' being applied to these curves is clear from Figure 2.20. This shows various sections through a pair of cones mounted with their vertices together and on a common axis.

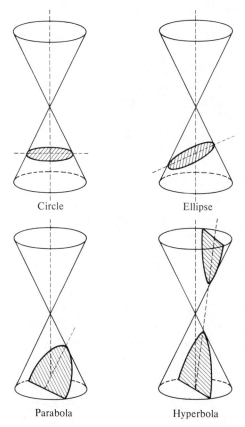

Circle   Ellipse

Parabola   Hyperbola

FIG. 2.20

Example. Sketch the graph of $y = 1/(x - 2)$.

This is similar to the previous example except that the singularity is now at $x = 2$ with a corresponding shift to the right by two units of the whole figure. When x is large, y is small. When x is less than 2, y is negative and, when x is greater than 2, y is positive. The graph is shown in Figure 2.21.

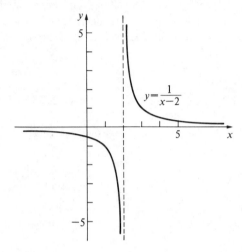

FIG. 2.21

Example. Assuming the thin lens formula,

$$\frac{1}{u} + \frac{1}{v} = \frac{1}{f},$$

where u is the object distance, v the image distance, and f the focal length, find an expression in terms of u for the separation s between a real object and a real image for a thin convex lens of focal length 10 cm. By drawing a graph of s against u find the minimum value of s and the value of u at which it occurs.

For a real object and a real image both u and v must be greater than the focal length and must be on opposite sides of the lens. A ray diagram is shown in Figure 2.22. Separation $s = u + v$ and so $v = s - u$. This can be substituted into the thin lens formula, together with $f = 10$ and solved for s.

$$\frac{1}{u} + \frac{1}{s-u} = \frac{1}{10},$$

$$10(s - u) + 10u = u(s - u).$$

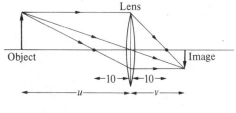

F IG . 2.22

Therefore,

$$s = \frac{u^2}{u - 10} = u + 10 + \frac{100}{u - 10}.$$

It is clear that there is a singularity at $u = 10$ and that as u becomes large $s \approx u + 10$, a straight line. Lines towards which a graph tends as one or other of the variables become large are known as *asymptotes*. The line $s = u + 10$ is an asymptote in this case and so is the vertical line $u = 10$. Asymptotes have occurred previously in the rectangular hyperbola $y = 1/x$ where the coordinate axes $x = 0$ and $y = 0$ are asymptotes. We are only interested in that part of the curve for which $u \geqslant 10$ (greater than or equal to 10) since there is no real image if $u < 10$ (less than 10). If u is just greater than 10, $u = 10 + \epsilon$ say, where ϵ is a small positive number, then

$$s = 10 + \epsilon + 10 + \frac{100}{\epsilon} = \frac{100}{\epsilon} + 20 + \epsilon.$$

Clearly if ϵ is very small and positive, s is very large and positive and, if ϵ increases, s decreases to start with. Also as u becomes large the graph starts to look like the line $s = u + 10$ which slopes upwards at $45°$. The graph of s against u therefore starts very high near $u = 10$, then falls, but must turn and eventually run upwards at $45°$. A table of values helps complete the picture.

u	11	15	20	25	30	40	50
s	121	45	40	$41\frac{2}{3}$	45	$53\frac{1}{3}$	$62\frac{1}{2}$

The graph is shown in Figure 2.23 from which it is seen that a minimum separation of about 40 cm occurs when $u = 20$ cm.

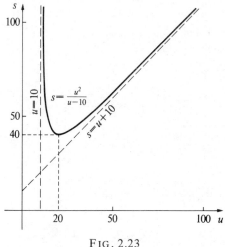

FIG. 2.23

Example. The speed S (cm s^{-1}) with which the sartorius of a frog can contract under a load L (g) is found to obey the empirical law

$$S = 0.95\left(\frac{70 - L}{L + 12}\right).$$

Discuss the main features of the graph of S against L. Sketch the section of the graph which lies in the first quadrant and explain the physical significance of the points at which it crosses the axes.

There is a singularity at $L = -12$ and the graph has a vertical asymptote there. As L becomes very large positive or negative S tends to -0.95. There is therefore a horizontal asymptote at $S = -0.95$. The graph is a rectangular hyperbola centred on these two asymptotes.

A sketch of that section of the graph which lies in the first quadrant is shown in Figure 2.24. It crosses the horizontal axis at $L = 70$. Since the speed is zero for this load of 70 g it can be interpreted as the load which is just too heavy for the muscle to move. The point at which the graph crosses the vertical axis gives the speed under a zero load situation. This is the fastest speed at which the muscle can contract and is about 5.5 cm s^{-1}.

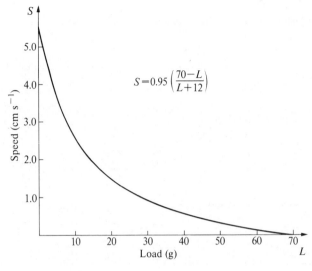

$$S = 0.95 \left(\frac{70-L}{L+12} \right)$$

F ig. 2.24

Exercises

1. Sketch the graphs of the following rational functions.

(a) $y = \dfrac{1}{x+1}$;   (b) $y = \dfrac{2}{x-1}$;   (c) $y = \dfrac{x}{x-1}$;

(d) $y = \dfrac{x-1}{x+1}$;   (e) $y = \dfrac{x}{x^2-1}$;   (f) $y = \dfrac{x^2}{x-1}$.

2.   A rectangular metal box without a lid is to be made with a square base and is to have a volume of 4 m^3. Find an expression for the area of metal required to make such a box in terms of a base dimension. By plotting this area against the base dimension find the least amount of metal required to make the box and its dimensions in this case.

3. Assuming the thin lens formula,

$$\frac{1}{u} + \frac{1}{v} = \frac{1}{f},$$

relating the object distance u, the image distance v, and the focal length f, sketch the variation of v against u for

(a) a convex lens of $f = 10$ cm;
(b) a concave lens of $f = -10$ cm.

4. The Michaelis–Menton equation,

$$\frac{r}{R} = \frac{c}{(k_M + c)},$$

expresses the initial rate r of an enzyme catalysed reaction as a fraction of the maximum rate R in terms of the concentration c of the substrate. k_M is a constant which is characteristic of each reaction and is known as the Michaelis constant. Sketch the typical shape of the curve produced by plotting r/R against c. What is the significance of k_M?

2.6. Exponential functions

An *exponential function* is a function in which the variable or a function of the variable occurs in the *exponent*, *power*, or *index*, each of these three words being equivalent. Since the simplest function of x is the linear function $mx + c$ with $m = 1$ and $c = 0$ the simplest exponential function has the form

$$a^x.$$

The number a is a positive constant, not equal to 1, known as the *base* of the exponential function. The reason for this value being excluded is that 1 raised to any power is still 1. An exponential function with this base would be a constant and not depend upon the exponent at all. Zero is excluded since zero raised to any power is either zero or infinity depending upon whether the power is positive or negative. Negative bases are excluded because the interpretation of an exponential function with a negative base in terms of conventional numbers is not possible. The same kind of problems arise with the solution of a quadratic equation which has a negative discriminant (cf. §2.3).

Exponential functions occur frequently in nature and are therefore important to the biologist. Two simple and no doubt familiar examples are the decay of a radioactive isotope and the unrestricted growth of a biological population. In the first case there is what is termed *exponential decay* and in the second *exponential growth*.

To illustrate the latter consider a situation in which at some reference time a population is of size P_0. If the birth rate exceeds the death rate the population will grow. Provided there are no restrictions such as shortage of food or space this (by now larger) population will grow even faster because there are even more individuals capable of reproduction. At the level of the individual cell what is happening is perhaps even clearer. If this particular type of cell divides after say half an hour then after half an hour for each original cell there will now be two cells. Both of these cells behave as the original cell and divide after a further half hour. There are now four cells and half an hour later there are eight and so on. At this rate the population doubles every half an hour. If deaths are also included the period for the doubling of the population will be somewhat larger but for the sake of simplicity we shall assume that the population does double every half an hour. If the population is originally P_0 its subsequent size can be tabulated.

t	0	$\frac{1}{2}$	1	$1\frac{1}{2}$	2	$2\frac{1}{2}$
P	P_0	$2P_0$	$4P_0$	$8P_0$	$16P_0$	$32P_0$

Note that at any particular time the population half an hour *earlier* is half the current size. We can therefore extrapolate backwards on the assumption that the particular growth pattern still holds and see that half an hour before the reference time $t = 0$ the population was $\frac{1}{2}P_0$ and half an hour before that it was $\frac{1}{4}P_0$. A graph of population P against time t is shown in Figure 2.25. The curve is asymptotic to the negative t-axis.

The mathematical relationship between P and t is not too difficult

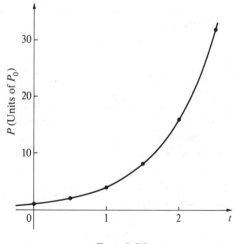

F IG. 2.25

to guess at. We require a relationship which introduces an extra factor of 2 every time t increases by $\frac{1}{2}$. Such a relationship is

$$P = P_0 2^{2t}.$$

This is clearly an exponential function with base 2 and exponent $2t$. Checking with various values of t shows that it does reproduce the table of values above. It is worth noting that since $2^{2t} = 2^{2 \times t} = (2^2)^t = 4^t$ (cf. Appendix A.2) the population can be written

$$P = P_0 4^t.$$

This is again an exponential relationship but now has base 4 and exponent t. It seems possible therefore that the base of an exponential function can be changed. An appropriate change must be made to the exponent to ensure that the value of the function for a particular value of the variable is unaltered. Putting $t = 2$ in 2^{2t} gives $2^4 = 16$ and in 4^t gives $4^2 = 16$. Putting $t = \frac{1}{2}$ in 2^{2t} gives $2^1 = 2$ and in 4^t gives $4^{\frac{1}{2}}$ which is interpreted as $\sqrt{4} = 2$ (cf. Appendix A.2). The graph (Figure 2.25) of this equation is smoothly rising from left to right and exists for all values of time t, both positive and negative. There is a unique interpretation for every exponent of the base whether this exponent is negative, positive, fractional, decimal, or even irrational such as $\sqrt{2}$ or transcendental like π.

We have seen one example in which the base can be changed, and it can be shown that any exponential may be written in terms of any base. This means that we can convert any exponential in any base to an exponential in a more conventional base if we wish to do so. There is one base which is usually taken as standard and it is the number e = 2.71828 . . . . This, like π = 3.141592 . . . is a never-ending decimal and, also like π, arises naturally. The reason for such a peculiar choice of base is given later in §5.4. There are other bases in fairly common use for one reason or another. One such is 10, since we count using tens, and another is 2 since doubling is a useful concept and computers count in twos. The particular exponential function e^x, sometimes written $\exp(x)$, is known as *the exponential function*. The graph of $y = e^x$ is shown in Figure 2.26 and it can be seen that as x increases y climbs steadily through positive values. Hence, if a is any positive number, a value k can be found such that

$$a = e^k.$$

This is shown on the graph. The number a is chosen on the y-axis and

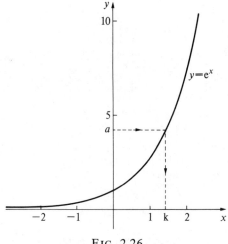

F IG. 2.26

using the curve $y = e^x$ a corresponding value of k can be found on the x-axis. This means that the exponential function a^x can be converted quite easily to base e since

$$y = a^x = (e^k)^x = e^{kx}.$$

The new exponent is kx. The number a can be chosen quite arbitrarily and this shows that any base a can be converted to base e with an appropriate change in the exponent.

Because exponentials with base e are so important most sets of mathematical tables contain tables of e^x and e^{-x}. It should be noted that, by the rules for indices (Appendix A.2)

$$e^{-x} = \frac{1}{e^x},$$

and so tables of e^x and tables of reciprocals would be sufficient. The reader should perhaps confirm correct use of the tables by looking up the following exponentials and checking that the given numbers are obtained. Tables usually contain values of x from 0 to 10. To obtain exponentials outside this range scientific notation (Appendix A.2) can be used with the index rules.

x	e^x	e^{-x}
1	2.7183	0.36788
2	7.3891	0.13534
3	20.086	0.049787
5	148.41	0.0067379
2.5	12.182	0.082085
4.3	73.680	0.013569
3.74	42.098	0.023754
1.627	5.0886	0.19652

Example. Find $e^{23.1}$.

$$e^{23.1} = e^{2.31 \times 10} = (e^{2.31})^{10} = (10.074)^{10}.$$

This final calculation can be performed using logarithms (cf. §2.7).

There are cases in which the standard base e may be of doubtful value. One example is the population equation

$$P = P_0 2^{2t}$$

produced earlier. This is concise and there is no doubt that the population doubles each time t increases by $\frac{1}{2}$. Written in terms of the base e it becomes

$$P = P_0 e^{0.69315 \times 2t} = P_0 e^{1.38629t}$$

since $2 = e^{0.69315}$. This form is much less neat and convenient.

If we now return to the more general exponential a^x some statements can be made about its behaviour. All graphs of $y = a^x$ pass through $(0, 1)$ since for any number a, $a^0 = 1$. If $a > 1$ (greater than 1), the function increases continuously with increasing x and rises rapidly once $x > 0$.

Example. $y = 2^x$.

x	−5	−2	−1	0	1	2	5
y	0.03125	0.25	0.5	1	2	4	32

If $0 < a < 1$, the function decreases continuously with increasing x and is asymptotic towards the positive x-axis.

Example. $y = (\frac{1}{2})^x$.

x	-5	-2	1	0	1	2	5
y	32	4	2	1	0.5	0.25	0.03125

Note that $\frac{1}{2} = 2^{-1}$ and so $y = (\frac{1}{2})^x = (2^{-1})^x = 2^{-x}$. Its graph is the mirror image in the y-axis of $y = 2^x$. Finally it should be noted that all exponential functions are always positive. Their graphs never quite touch the x-axis and certainly never fall below it. Graphs of some exponential functions are shown in Figure 2.27.

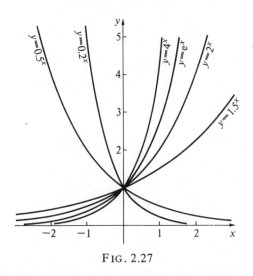

F IG. 2.27

Example. The build up of charge Q on a nerve with respect to time t may be represented by an equation of the form

$$Q = Q_0(1 - e^{-kt})$$

where Q_0 is a constant and k is a positive constant. Sketch the general shape of the graph of this relationship and explain the significance of Q_0.

The function e^{-kt} is a negative exponential and decreases from unity towards zero as t increases. Its shape is characteristic of exponential decay and is like the graphs of $y = (0.5)^x = 2^{-x} = e^{-0.693x}$ and $y = (0.2)^x = 5^{-x} = e^{-1.609x}$ shown in Figure 2.27. The larger the value of

k the steeper will be the initial descent. A typical graph of $y = e^{-kt}$ is shown in Figure 2.28a together with the corresponding graph of $y = (1 - e^{-kt})$, which rises from zero towards unity. The latter is a mirror image of the former in the horizontal line $y = \frac{1}{2}$.

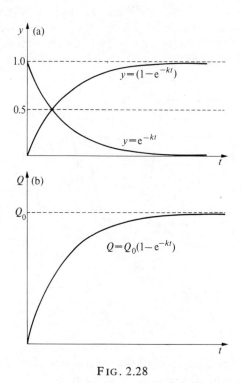

F IG . 2.28

The graph of $Q = Q_0(1 - e^{-kt})$ will look like the graph of $y = (1 - e^{-kt})$ with its vertical axis scaled by the factor Q_0. It rises from $Q = 0$ at $t = 0$ (the origin) and levels off as $Q = Q_0$ is approached from below as shown in Figure 2.28b. Hence Q_0 is the charge finally reached by the nerve if the build up were allowed to proceed undisturbed for an infinite time.

Exercises

1. An organism with an initial mass of 0.5 g doubles its mass every $2\frac{1}{2}$ hours. Find an expression for the mass as a function of time and hence calculate the mass after 24 hours.

2. It is said that an Eastern prince had so enjoyed playing chess that he offered its inventor any reward the man chose. The man's choice was simple. He asked that he be given a chess-board and allocated one grain of rice for the first square, two grains for the second, four for the third, eight for the fourth, and so on until all 64 squares had been allocated rice. The prince and his court laughed at this simple request which was immediately granted. Who should have been laughing and why? If it is assumed that the average mass of a grain of rice is 0.02 g, find approximately the mass allocated to the last square and the total mass of rice allocated to all squares. Compare this last result with,
(a) the annual world production of rice, $(4 \times 10^8$ tonnes);
(b) the mass of the Earth $(6 \times 10^{21}$ tonnes).

3. For values of x from 0 to 10 draw graphs of x^2, x^3 and 2^x against x. As x increases which eventually becomes the largest function?

4. Using an accurate graph of the type shown in Figure 2.26 find the value of k which enables $P = P_0 2^t$ to be written in the form $P = P_0 e^{kt}$.

5. A radioactive isotope decays at a rate proportional to its current mass. Its half-life T is the period over which its mass halves. If initially its mass is m_0, then its mass m at any subsequent time t is given by $m = m_0(\frac{1}{2})^{t/T}$. Find the value of k which enables this formula to be written as

$$m = m_0 e^{-kt/T}.$$

6. If initially there is 1 g each of carbon 14 (half-life 5570 years), cobalt 60 (5.3 years), phosphorus 32 (14.3 days), and sodium 24 (15 hours) how much of each will remain after 1 day, 100 days, 30 years, and 3000 years?

7. The temperature of a body is given by

$$T = 30 + 70e^{-0.04t}$$

where T is in $°C$ and t is the time elapsed in minutes. Sketch a graph of T against t and calculate the time it takes to cool to $70°C$.

8. Evaluate

$$\left(1 + \frac{1}{n}\right)^n$$

for $n = 1, 2, 5, 10, 100, 10\ 000, 1\ 000\ 000$. The following logarithms may be required.

$$\log (1.01) \quad\quad = 4.3214 \times 10^{-3},$$
$$\log (1.0001) \quad = 4.3427 \times 10^{-5},$$
$$\log (1.000001) = 4.3429 \times 10^{-7}.$$

Do these results give an indication of the possible value of $(1 + (1/n))^n$ as n tends to infinity?

2.7. Logarithmic functions

Logarithmic functions are very closely related to the exponential functions of the previous section. Both have a *base* defined in the same way and indeed the logarithm of a number is defined in terms of a power, index, or exponent.

The *logarithm to the base a* of a number x is the power to which a must be raised for the resulting number to be equal to x.

This definition implies that if a number y can be found such that

$$a^y = x$$

then y is the logarithm to the base a of x. This is written

$$y = \log_a x.$$

Since a^y is always positive for a carefully defined base a ($a > 0, a \neq 1$) and any number y then x is always positive. It is not possible to have the logarithm of a negative number.

The most frequently encountered logarithms are those to the base 10 and are known as *common logarithms*. They are described by the symbols log or lg. It is becasue they are so common that the base is assumed to be 10 if not stated. By the rules for indices or exponents (Appendix A.2) two numbers expressed as indices of a common base can be multiplied together by adding their indices or exponents. Hence,

$$3^2 \times 3^{4.5} = 3^{6.5}.$$

Logarithms provide a way of expressing all numbers in terms of a common base and hence facilitating multiplication and division by reducing it to addition and subtraction. For common logarithms this base is 10 and indeed tables of common logarithms are simply tables of indices to which 10 must be raised to equal a given number.

Example. Multiply 2 by 3 using common logarithms.

$$\log 2 = 0.30103 \quad \text{means} \quad 2 = 10^{0.30103},$$
$$\log 3 = 0.47712 \quad \text{means} \quad 3 = 10^{0.47712}.$$

Hence,

$$2 \times 3 = 10^{0.30103} \times 10^{0.47712} = 10^{(0.30103+0.47712)} = 10^{0.77815}.$$

The product is recovered from tables of common antilogarithms which are simply exponentials with base 10. Since a common base is being used this base is omitted and the working reduces to

No.	Log
2	0.30103
×3	0.47712 +
6	0.77815

Example. Divide 7 by 2 using common logarithms.

$$\log 7 = 0.84510 \quad \text{means} \quad 7 = 10^{0.84510},$$
$$\log 2 = 0.30103 \quad \text{means} \quad 2 = 10^{0.30103}.$$

Hence,

$$7 \div 2 = 10^{0.84510} \div 10^{0.30103} = 10^{(0.84510-0.30103)} = 10^{0.54407}$$

or, more concisely

No.	Log
7	0.84510
÷2	0.30103 −
3.5	0.54407

The common logarithms of all numbers between 1 and 10 lie between 0 and 1. For numbers outside this range scientific notation (Appendix A.2) can be very useful.

Example. Find the common logarithm of 172.

Now, $172 = 1.72 \times 10^2$ and since $\log 1.72 = 0.23553$ we have
$$172 = 10^{0.23553} \times 10^2 = 10^{2.23553}.$$

Hence,

$$\log 172 = 2.23553.$$

Example. Find the common logarithm of 0.000397.

$$0.000397 = 3.97 \times 10^{-4}$$
$$\log 3.97 = 0.59879.$$

Therefore,

$$0.000397 = 10^{0.59879} \times 10^{-4} = 10^{(0.59879 - 4)} = 10^{-3.40121}.$$

Hence

$$\log(0.000397) = -3.40121.$$

However, it is not conventional to write logarithms in this form. It is more usual to keep the decimal part or *mantissa* a positive number and adjust the whole number part or *characteristic* accordingly. The characteristic is precisely the power of 10 which occurs in the scientific notation form of the original number. If it is negative a minus sign or *bar* is written above it. The conventional form is therefore

$$\log(0.000397) = \bar{4}.59879,$$

meaning

$$-4 + 0.59879.$$

Manipulation of these bar numbers is fairly straightforward if it is remembered that the bar is simply a minus sign and only applies to the whole number part of the logarithm. The decimal part is kept positive.

Example. Use common logarithms to find $\sqrt[3]{0.3674}$.

Now, $0.3674 = 3.674 \times 10^{-1} = 10^{0.56514} \times 10^{-1} = 10^{\bar{1}.56514}$. Therefore,

$$\sqrt[3]{0.3674} = (0.3674)^{\frac{1}{3}} = (10^{\bar{1}.56514})^{\frac{1}{3}}.$$

We now wish to multiply $\bar{1}.56514$ by $\frac{1}{3}$ keeping the decimal part positive. This can be done by writing $\bar{1}.56514$ as $\bar{3} + 2.56514$ and then dividing by 3. Therefore,

$$\sqrt[3]{0.3674} = (10^{\bar{3}+2.56514})^{\frac{1}{3}} = 10^{\bar{1}.85505}.$$

The whole process can be shortened to

No.	Log
0.3674	$\bar{1}.56514$
	$\bar{3} + 2.56514$
	$\div 3$
7.162×10^{-1}	$\bar{1}.85505$

Apart from common logarithms, which have base 10, there are logarithms with other bases which are used extensively. *Natural logarithms* have the number e as their base and are denoted by $\log_e$ or ln. They are sometimes referred to as hyperbolic logarithms or Naperian logarithms (after the Scottish mathematician John Napier, 1550-1617). Natural logarithms are more difficult to use than common logarithms because the base e does not fit in well with our decimal system of counting. Whereas, using common logarithms, $\log 1 = 0, \log 10 = 1, \log 100 = 2,$ $\log 1000 = 3,$ etc., the use of natural logarithms gives $\ln 1 = 0, \ln 10 =$ 2.30259, $\ln 100 = 4.60517,$ and $\ln 1000 = 6.90776.$ It is perhaps unfortunate, therefore, that the number e, like π, has such an important part to play in the world around us. The situations in which it arises lead to a need for the exponential function e^x and the associated natural logarithmic function $\ln x$.

Most tables give natural logarithms of numbers from 1 to 10 as for common logarithms. To find the natural logarithm of a number outside this range scientific notation is used and the natural logarithm of the appropriate power of 10 is added to the logarithm of the significant figures. To facilitate this the natural logarithms of positive and negative powers of 10, $\ln 10^n$ and $\ln 10^{-n}$ respectively, are often tabulated at the bottom of the pages of the natural logarithms of the numbers 1 to 10.

Example. Find ln 23 from tables.

Using scientific notation $23 = 2.3 \times 10$ and so

$$\ln 23 = \ln(2.3 \times 10) = \ln 2.3 + \ln 10$$

$$= 0.83291 + 2.30259$$

$$= 3.13550.$$

Example. Find $\ln 0.035$ from tables.

$$\ln 0.035 = \ln(3.5 \times 10^{-2}) = \ln 3.5 + \ln 10^{-2}$$

$$= 1.25276 + \bar{5}.39483$$

$$= \bar{4}.64759.$$

Mathematical tables do not contain anti-natural logarithms since tables of exponentials e^x do precisely this job. The anti-natural logarithm of a number x *is* e^x since, by the definition of a logarithm, if $\ln y = x$ then $y = e^x$.

Although not as common as logarithms to the base 10 or e, logarithms to the base 2 are not unknown. They are used in calculations associated with cell division for example where the logarithm to the base 2 of the number of cells at any time is equal to the number of generations through which the population has passed. This assumes that generations do not overlap and that the population originated from a single cell.

A number of properties can be established directly from the properties of indices (Appendix A.2) and apply to logarithms of any base.

1. $$\log_a 1 = 0.$$

2. $$\log_a a = 1.$$

3. $$\log_a x + \log_a y = \log_a(xy).$$

4. $$\log_a\left(\frac{1}{x}\right) = -\log_a x.$$

5. $$\log_a x - \log_a y = \log_a\left(\frac{x}{y}\right).$$

6. $$\log_a(x^n) = n \log_a x.$$

7.
$$\log_b a = \frac{1}{\log_a b}.$$

8.
$$\log_b x = \frac{\log_a x}{\log_a b}.$$

9.
$$\log_a(a^x) = x.$$

10.
$$a^{\log_a x} = x.$$

Example. To establish that $\log_b x = \log_a x/\log_a b$ for any positive number x and any bases a and b.

Let $\log_b x = L$ so that $x = b^L$ by the definition of logarithms. Similarly let $\log_a x = M$ and $\log_a b = N$ so that $x = a^M$ and $b = a^N$. Hence we can write

$$x = b^L = (a^N)^L = a^{LN}.$$

However, $x = a^M$ and so $a^M = a^{LN}$. The shape of an exponential graph is such that it is falling steadily if $0 < a < 1$ or rising steadily if $a > 1$ and so the only conclusion which can be drawn from $a^M = a^{LN}$ is that

$$M = LN \quad \text{or} \quad L = \frac{M}{N}$$

i.e.
$$\log_b x = \frac{\log_a x}{\log_a b}.$$

Property 10 is interesting and very useful. Together with property 9 it demonstrates that an exponential and a logarithmic function with the same base are inverse functions with respect to each other, that is, they cancel each other out. It can be used to convert an exponential with one base to an exponential with any other base.

Example. Convert a^x to base b.

Using property 10, $a = b^{\log_b a}$ and so $a^x = (b^{\log_b a})^x = b^{x \log_b a}$ and the new exponent must be $x \log_b a$ in order to preserve the equality of these two forms.

The general shape of the graph of

$$y = \log_a x,$$

where $a > 1$, is shown in Figure 2.29. The number x must be positive and the graph will always pass through $(1, 0)$ and $(a, 1)$ because of properties 1 and 2.

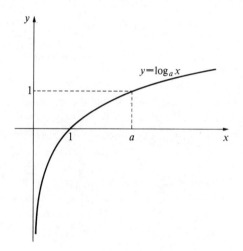

FIG. 2.29

Example. Changes in sound intensity levels are measured in decibels (dB). The gain or loss is defined by

$$10 \log\left(\frac{I}{I_0}\right) \quad \text{dB}$$

where I_0 is the initial or base intensity and I is the final intensity. In SI units the arbitrary standard base intensity is taken to be 10^{-12} watts per square metre (Wm^{-2}) and is known as the subjective threshold intensity. A laboratory centrifuge produces a sound level of 64 dB at a distance of 1 metre. Find the intensity of sound it is producing at this distance. If a second similar centrifuge starts up at a distance of 1 metre so that the sound intensity doubles find the new sound level in decibels. Sound insulating covers which are claimed to reduce sound by

45 dB are placed over the centrifuges. Find the intensity of sound to be expected at 1 metre in this case and the new sound level in dB.

A level of 64 dB has an intensity of I Wm^{-2} where

$$64 = 10 \log \left(\frac{I}{I_0} \right)$$

and $I_0 = 10^{-12}$ Wm^{-2}. Hence

$$\frac{I}{I_0} = 10^{6.4}$$

and so

$$I = I_0 \times 10^{6.4} = 10^{-12} \times 10^{6.4} = 10^{-5.6} = 2.512 \times 10^{-6} \text{ Wm}^{-2}.$$

If the intensity doubles to 5.024×10^{-6} Wm^{-2} then the new level is

$$10 \log \left(\frac{5.024 \times 10^{-6}}{10^{-12}} \right) = 67 \text{ dB}.$$

If the covers reduce the sound by 45 dB then the expected intensity of sound I is given by

$$10 \log \left(\frac{I}{5.024 \times 10^{-6}} \right) = -45.$$

Therefore

$$I = 5.024 \times 10^{-6} \times 10^{-4.5}$$

$$= 1.589 \times 10^{-10} \text{ Wm}^{-2}.$$

The new level is

$$10 \log \left(\frac{1.589 \times 10^{-10}}{10^{-12}} \right) = 22 \text{ dB}.$$

Note that this last result could have been obtained by subtracting the claimed reduction of 45 dB from the total sound level of 67 dB.

Exercises

1. Find the common logarithms of

(a) 2.67;   (b) 6.932;   (c) 0.3718;

(d) 0.0271;   (e) 0.000361;   (f) 13.62;

(g) 2.8×10^4;   (h) 6.71×10^{-6}.

2. Use common logarithms to evaluate

(a) 10.62×1.478;   (b) 123.4×0.1234;

(c) $2.671 \div 93.86$;   (d) $0.6298 \div 49\,320$.

3. Use common logarithms to evaluate

(a) 2^4;   (b) $(10.7)^2$;   (c) $(4.28)^{15}$;

(d) $(1.08)^{20}$;   (e) $(0.2974)^3$;   (f) $\sqrt{3.682}$;

(g) $\sqrt{0.269}$;   (h) $(0.0582)^{1.6}$;   (i) $\sqrt[5]{987.4}$;

(j) $(0.00657)^{\frac{1}{27}}$.

4. Use tables of natural logarithms to find

(a) $\ln(2.7)$;   (b) $\ln(6.382)$;   (c) $\ln(15.62)$;

(d) $\ln(0.00638)$;   (e) $\ln(49\,630)$.

5. Use natural logarithms to find the answers to questions 2 and 3.

6. On the same graph paper draw the graphs of $y = \log x$ and $y = \ln x$ for values of x from 0.1 to 10.

7. Use the rules for indices (Appendix A.2) to prove the 10 properties of logarithms listed earlier.

8. Use the property

$$\log_b x = \frac{\log_a x}{\log_a b}$$

to find

(a) $\log_2 3$;   (b) $\log_3 2$;   (c) $\log_2 64$;   (d) $\log_5 15$;   (e) $\log_2(1.5)$.

9. Find $\ln x$ in terms of $\log x$.

10. In each of the following cases find x as a function of y.

(a) $y = 10^x$;   (b) $y = e^x$;   (c) $y = 4e^{5x}$;

(d) $y = 2 + e^{x+3}$;   (e) $y = \frac{1}{2}(e^x + e^{-x})$.

11. The speed of a signal transmitted along a coaxial cable is found to be proportional to $-x^2 \ln x$ where x is the ratio of the radius of the inner conductor to that of the outside of the insulating sheath $(0 < x < 1)$. If it is assumed that a nerve behaves in this manner, find, by drawing an appropriate graph, the ratio of the radii of axon to myelin sheath which produces the maximum speed of impulse transmission.

2.8. Trigonometric functions

Before considering these functions something should be said about *angles* and the ways in which they are described. The use of angles provides a method of comparing the direction of a line with that of a fixed line or describing the relative directions of a pair of intersecting lines. A pair of perpendicular lines are said to be at right angles to one another for instance. Finer division than this is required and on one system of measurement the degree symbol $^\circ$, is used. A *degree* is one three hundred and sixtieth part of a circle and so a right angle consists of 90°. The degree can be subdivided into 60 minutes $(60')$ and each of these into 60 seconds $(60'')$. Another system of measurement uses radians, symbol rad, and in many calculations is of more use than angles measured in degrees. An angle in *radians* is defined to be the length of the circular arc it subtends divided by the radius of the circle. In Figure 2.30 the circular arc has length l, the radius of the circle is r, and so the angle ϕ in radians is given by

$$\phi = \frac{l}{r}.$$

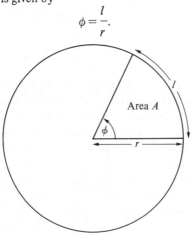

FIG. 2.30

A complete circle has an arc length of $2\pi r$, its radius being r, and so a complete circle subtends 2π rad. Hence,

$$360° \equiv 2\pi \text{ rad},$$

$$90° \equiv \frac{\pi}{2} \text{ rad}.$$

To convert from degrees to radians multiply by $2\pi/360$ (or $\pi/180$). To convert from radians to degrees multiply by $360/2\pi$ (or $180/\pi$).

One radian is approximately $57°$.

Note that since an angle in radians is defined to be the ratio of two lengths it has no dimensions. Two immediate uses of an angle in radians occur directly from the definition. Rearrangement of the equation shows that the arc length l is given by

$$l = r\phi,$$

provided ϕ is measured in radians. The area of a sector of a circle is proportional to the arc length of that sector which in turn is proportional to the angle at the centre. Consider a sector of area A of a circle of radius r (Figure 2.30). The arc length is l. The ratio that A bears to the area of the whole circle is the same as the ratio that l bears to its circumference. Hence

$$\frac{A}{\pi r^2} = \frac{l}{2\pi r} = \frac{r\phi}{2\pi r}.$$

Therefore,

$$A = \tfrac{1}{2}r^2\phi,$$

provided ϕ is in radians. These two examples show that radians have their uses and can indeed produce very neat formulae.

Trigonometric functions of angles between $0°$ and $90°$ (0 and $\pi/2$ rad) can be defined in terms of the ratios that the lengths of the sides of a right-angled triangle bear to one another when referred to a specific angle. Such a triangle is shown in Figure 2.31 and the three fundamental trigonometric functions are defined as follows.

Sine of ϕ, written $\sin \phi$, is equal to $\dfrac{\text{Opposite}}{\text{Hypotenuse}} = \dfrac{b}{c}$.

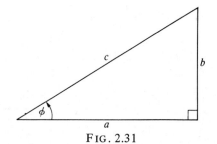

F IG . 2.31

Cosine of ϕ, written cos ϕ, is equal to $\dfrac{\text{Adjacent}}{\text{Hypotenuse}} = \dfrac{a}{c}$.

Tangent of ϕ, written tan ϕ, is equal to $\dfrac{\text{Opposite}}{\text{Adjacent}} = \dfrac{b}{a}$.

From these definitions it is immediately obvious that

$$\tan \phi = \frac{\sin \phi}{\cos \phi}.$$

Since the triangle is a right-angled triangle, by Pythagoras,

$$a^2 + b^2 = c^2.$$

Hence,

$$\left(\frac{a}{c}\right)^2 + \left(\frac{b}{c}\right)^2 = 1.$$

Therefore,

$$\cos^2 \phi + \sin^2 \phi = 1.$$

The reference to a right-angled triangle limits these definitions to angles between $0°$ and $90°$ (0 and $\pi/2$ rad). A more general set of definitions can be formulated using a circle of unit radius with its centre at the origin of a Cartesian coordinate system (Figure 2.32). A radius is drawn at an angle ϕ measured in an anti-clockwise (positive) direction from the x-axis. Since the radius is 1 the trigonometric functions are defined by

$$\cos \phi = x, \quad \sin \phi = y, \quad \tan \phi = \frac{y}{x}.$$

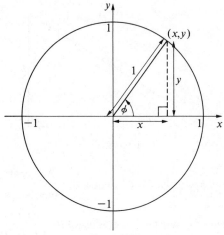

FIG. 2.32

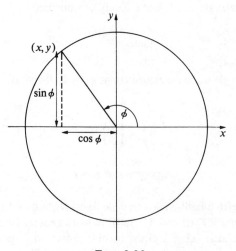

FIG. 2.33

Using this figure, angles greater than one right angle can be considered. If the radius lies in the second quadrant (Figure 2.33) so that ϕ lies between one and two right angles, the same relationships hold for the trigonometric functions. In this case x is negative and y is still positive and so cos ϕ is negative, sin ϕ positive, and tan ϕ negative. The definitions apply to all angles including angles greater than 360° (2π rad)

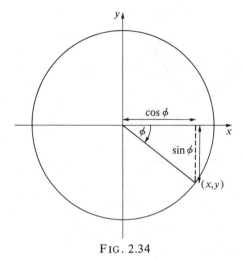

F IG. 2.34

and negative angles (Figure 2.34). An angle is negative if it is measured clockwise from the *x*-axis.

The graphs of sin ϕ, cos ϕ, and tan ϕ can be drawn from the unit circle representation of the trigonometric functions and are shown in Figure 2.35. The tangent graph is the most complex since there are singularities (cf. §2.5) wherever cos ϕ = 0. They are all *periodic*, that is they repeat themselves after a fixed interval or *period*. The period of sin ϕ and cos ϕ is 360° (2π rad) and of tan ϕ is 180° (π rad). It is their periodicity that makes trigonometric functions so important. The fact that they are periodic should not be so surprising since we have used a swinging radius to define them. This radius reaches its starting point every 360° (2π rad) and then swings round over exactly the same path again. We can therefore expect any phenomenon caused directly or indirectly by a regular rotation to be periodic and to involve or require the use of trigonometric functions for its description. Such obvious examples as the Earth spinning about its axis and in turn moving round the sun come to mind. Each produces all kinds of biological phenomena from the daily opening and closing of flowers to the seasonal migrations of birds and animals. Taken even further we might expect the description of any periodic phenomenon to require the use of trigonometric functions, whether the periodicity is in space or time. The use of trigonometric functions may range from proteins to brain waves.

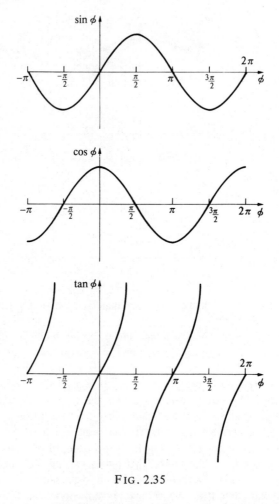

FIG. 2.35

Example. A population of small birds within a specified area is found to vary over a year between about 1000 and 1500. The minimum occurs at the end of March and the maximum some six months later. If the variation is assumed to be sinusoidal find an expression for the approximate population P as a function of time t (days) from the beginning of the year.

A sinusoidal variation takes place about a mean population size of 1250. The size or *amplitude* of this variation is 250 above and below the mean value. If the total period is one year and a minimum occurs after three months then at the beginning of the year the population must be at its mean value. It then decreases for three months, increases for six months and decreases for three months to end the year where it started. Such behaviour can be produced by subtracting a sine wave (Figure 2.35) of amplitude 250 from the mean population of 1250. Initially nothing is subtracted but after one quarter of a period the sine wave has risen to its maximum and so at three months 250 is subtracted. This gives the minimum at the end of March. The angle at this point is $\pi/2$ or one quarter of 2π. In terms of days it is $2\pi \times (30/365)$ approximately. The population can therefore be represented by

$$p = 1250 - 250 \sin\left(\frac{2\pi t}{365}\right) = 250\left(5 - \sin\left(\frac{2\pi t}{365}\right)\right).$$

A further use of trigonometric functions is in the resolution of forces where the effect of a force in a particular direction may need to be calculated. In such cases it is Pythagoras' theorem and the result $\sin^2\phi + \cos^2\phi = 1$ rather than periodicity which is of importance. Such problems occur when describing the action and leverage of muscles. As an example consider the action of a muscle, biceps say, in enabling an arm to support a weight. A simplified diagram is shown in Figure 2.36. The biceps exert a force F along its length, a direction which is not perpendicular to the forearm. In order to take moments about the elbow we wish to know the amount or *component* of the force in the vertical

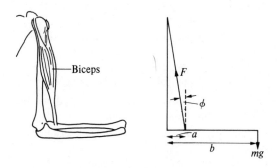

FIG. 2.36

direction. This component is $F \cos \phi$ and so, since the anti-clockwise moment must equal the clockwise moment,

$$(F \cos \phi)a = (mg)b.$$

If the lengths a and b are known, together with the mass m of the weight and the angle ϕ, then the force F can be calculated.

The tangent of an angle has been used implicitly in §2.2 on linear functions. It may be recalled that the slope m of a line was defined to be the increase in the dependent variable (y) resulting from a unit increase in the independent variable. Reference to Figure 2.37 will show

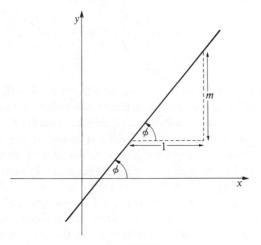

FIG. 2.37

that this is the tangent of the angle that the line makes with the horizontal. In other words, if ϕ is the angle that the line makes with the horizontal axis, measured in an anti-clockwise direction, then the slope m of the line is given by

$$m = \tan \phi.$$

Before concluding this section mention should be made of three other trigonometric functions which are simply the reciprocals of $\sin \phi$, $\cos \phi$, and $\tan \phi$. These are, respectively, the *cosecant, secant,* and *cotangent*.

$$\operatorname{cosec} \phi = \frac{1}{\sin \phi},$$

$$\sec \phi = \frac{1}{\sin \phi},$$

$$\cot \phi = \frac{1}{\tan \phi}.$$

Finally, the earlier work has almost brought us to the point of deriving the equation of a circle. Recalling that $\cos^2 \phi + \sin^2 \phi = 1$ and that on the unit circle we defined $\cos \phi = x$ and $\sin \phi = y$ we have

$$x^2 + y^2 = 1,$$

which is the equation of a circle with centre at the origin and radius 1. If the radius were a, the equation would have been

$$x^2 + y^2 = a^2.$$

The only conic section (cf. §2.5) whose equation has not been mentioned is the ellipse. For the sake of completeness it might be as well to state that the equation of an ellipse centred on the origin with semi-major axis of length a and semi-minor axis of length b is

$$\frac{x^2}{a^2} + \frac{y^2}{b^2} = 1.$$

Note that if $b = a$ this reduces to the equation of a circle of radius a. The circle and ellipse are shown in Figure 2.38.

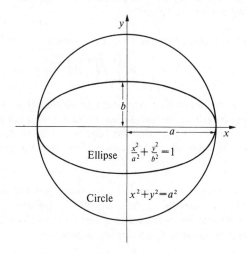

F IG . 2.38

Exercises

1. Express the following angles in radians.

(a) $45°$;   (b) $30°$;   (c) $135°$;   (d) $40°$;

(e) $80°$;   (f) $100°$;   (g) $200°$;   (h) $300°$;

(i) $-90°$;   (j) $-10°$.

2. Express the following angles in degrees.

(a) $\frac{1}{3}\pi$ rad;   (b) $\frac{1}{4}\pi$ rad;   (c) 0.1 rad;   (d) 0.25 rad;

(e) 0.5 rad;   (f) 1.5 rad;   (g) 2 rad;   (h) 6 rad;

(i) -0.3 rad;   (j) -2 rad.

3. Sketch the graphs of the following trigonometric equations.

(a) $y = \sin(2x)$;   (b) $y = \sin(\frac{1}{2}x)$;   (c) $y = \cos(3x)$;

(d) $y = 1 + \sin x$;   (e) $y = \sin^2 x \ (= (\sin x)^2)$;

(f) $y = \sin x + \sin(2x)$;   (g) $y = \tan(2x)$.

4. Draw graphs of sunrise and sunset (GMT) against the week of the year. The data can be obtained from most diaries. From these graphs deduce formulae which would give sunrise, sunset, and hence hours of daylight for any day of the year.

5. The angular deflection ϕ, in radians, of a simple pendulum is given by

$$\phi = \frac{\pi}{12} \sin \left\{ \sqrt{\left(\frac{g}{l}\right)} t \right\}$$

where t is time in seconds, g the local gravitational acceleration in m s^{-2}, and l the length of the pendulum in metres. If it is assumed that $g = 9.81$ m s^{-2} and $l = 1.5$ m, find the time at which the deflection is a maximum and the value of the maximum. Find also the period of the oscillations.

6. Periodic fluctuations in a predator or prey population are often accompanied by corresponding fluctuations in the prey or predator population. Some studies of lynxes in a certain region of Canada have shown that their population P_L can be represented by

$$P_L = 40\ 000 + 35\ 000 \sin \left(2\pi \frac{t}{T} \right)$$

where T, the periodic time, is 11 years and t is the time in years from some base date. Studies of the lynxes' primary prey, horseshoe hares, showed that their population also varied sinusoidally with a period of 11 years. It was found however that the hares reached a maximum of 110 000 two years before the lynxes reached their maximum. The minimum hare population was found to be 10 000. Find an equation for the hare population P_H as a function of time t. Draw graphs of P_L and P_H against t and hence determine when they are equal.

2.9. The function notation

In the previous sections various particular functions have been discussed. Occasions may arise when we wish to indicate that y is a function of x without specifying a particular functional relationship. This can be done by the use of the *function notation* in which, in its most straightforward form, the letter 'f' for function is used. We write

$$y = f(x).$$

Specific problems may require other variables. For example, we may wish to convey that body temperature T during periods of sleep is a function of time t. This can be indicated by

$$T = f(t).$$

It may be found that a particular parameter, time say, is the independent variable for more than one dependent variable. In this case letters following f in the alphabet are generally used to indicate these different dependences. Extending the above example it might be thought that, in addition to temperature, pulse rate P and stomach acidity A are functions of time during sleep. We therefore write

$$T = f(t),$$
$$P = g(t),$$
$$A = h(t).$$

In cases like this the introduction of the extra letters, f, g, and h, associated with T, P, and A respectively may cause confusion. The following alternative notation is sometimes used to prevent this. We write

$$T = T(t),$$
$$P = P(t),$$
$$A = A(t),$$

in which no extra letters are introduced. The notations $y = f(x)$ and $y = y(x)$ are equivalent. Both mean that y is a function of x.

It is sometimes useful to indicate that a function has been evaluated at a particular value of its *argument*, that is, the variable (or expression) inside it. To do this the particular value is written in. If we wished to indicate the value of $f(x)$ obtained when $x = a$ we write $f(a)$.

Example. If $f(x) = x + 1/x^2$ find $f(1), f(-1), f(2.7),$ and $f(\pi)$

$$f(1) \; = \; 1 + \frac{1}{1^2} = 2.$$

$$f(-1) \; = \; -1 + \frac{1}{(-1)^2} = 0.$$

$$f(2.7) \; = \; 2.7 + \frac{1}{(2.7)^2} = 2.837.$$

$$f(\pi) \; = \; \pi + \frac{1}{\pi^2} = 3.243.$$

Example. Find the simplest function $f(x)$ such that $f(2) = 5$ and $f(5) = 2$.

We require the simplest function which passes through the two points (2, 5) and (5, 2). Such a function is linear (a straight line) and so

$$f(x) = mx + c.$$

Using the result at the end of §2.2 the equation of the line is

$$\frac{y-5}{x-2} = \frac{2-5}{5-2} = -1.$$

Hence, $\qquad\qquad\qquad\qquad y = 7 - x.$

Therefore, $\qquad\qquad\qquad f(x) = 7 - x.$

Exercises

1. If $f(x) = x^2 - 2x - 3$, find

(a) $f(0)$;   (b) $f(3)$;   (c) $f(-1)$;   (d) $f(1)$;   (e) $f(-2)$.

2. Show that if $f(x) = A \sin x + B \cos x$, where A and B are constants, then $f(x + 2\pi) = f(x)$ for any value of x in radians. The trigonometric formulae in Appendix B.1 may be of value.

3 Experimentation

3.1. Empirical laws

EMPIRICAL LAWS are laws or relationships established between variables as a direct consequence of experimentation or observation. They are distinct from laws originally formulated as hypotheses from theory and subsequently checked and confirmed by experiment.

In order to establish empirical laws experimentation and/or observation is undertaken or has already taken place. In either case qualitative or quantitative data are obtained relating some phenomenon to a certain variable or variables. In the case of quantitative data tables of results are produced and graphs drawn so that the general features of the relationship can be established. Ideally it is hoped that a precise mathematical equation can be formulated and as an initial step in this direction a knowledge of various functions and their graphs can be invaluable. It may be possible to say immediately that the relationship is linear or that it seems to be exponential. Such informed guesses can then be checked by more analytical methods. Some of the simpler analytical methods are outlined in subsequent sections.

If a mathematical relationship can be found it is likely to be used for one or both of two purposes. The first of these is prediction. Experimental data is obtained at a finite number of points within a certain range of the independent variable. We might wish to predict the value of the dependent variable at intermediate points within this range (interpolation, cf. §3.7) or at points outside the range altogether (extrapolation, cf. §3.8). The former is usually quite safe since there are known experimental points on both sides of the point to be predicted which can be used as guides. Because it is an intermediate point, the same law can be assumed to apply as at the experimental points. The latter (extrapolation) must be applied with much more care. There are experimental points on one side only of the point to be predicted and the further away one gets from these points the less reliable are the predictions likely to be. The physical situation can change for instance. An empirical law might be established about the genetic effects of weak radiation on cells. It would be unreasonable to extrapolate such a law to the point of making predictions about high levels of radiation without some further experimental evidence to back them up. Indeed, if such evidence were available the process would cease to be one of extrapolation and become one of interpolations.

The second purpose for which an empirically established mathematical relationship might be used is to provide insight into the underlying biological processes being studied. It would be of great significance if a population grew according to a power law rather than to an exponential law as might have been expected. Some explanation of this divergence from the expected would have to be sought which might lead to the development of a new theory of growth for the population.

3.2. Dimensional analysis

It is perhaps stating the obvious to remark that the = sign in an equation means that the right-hand side is equal to the left-hand side. Because such a remark is obvious its significance may not be fully appreciated. There would be no question that something was amiss if the right-hand side came to 7.43 but the left-hand side to 4.26. As long as the numbers balance most people are satisfied that the equation is correct. The full meaning of the = sign has not been appreciated. We would not say that 4 metres were equal to 4 seconds even though the numbers were the same. An equation must balance both numerically *and* dimensionally. This fact can be used to check equations or even to help formulate them. The technique is known as *dimensional analysis*. It involves comparing the dimensions or units of measurement on each side of an equation. If they do not balance the equation is certainly incorrect. The converse however is not true. If they do balance the equation is *not necessarily* correct.

It may not be easy to compare the dimensions on each side of an equation because of the vast number of units employed in measuring physical quantities. When this is the case all units on both sides of the equation are converted to *basic units* which are limited in number and make comparisons easier. The basic units in the Système International (S.I.) are

Physical quantity	Unit	Symbol
length	metre	m
mass	kilogram	kg
time	second	s
electric current	ampere	A
temperature	kelvin	K
luminous intensity	candela	cd
amount of a substance	mole	mol

It is a general principle that units which are named after people have symbols with capital letters. In spite of this the words Ampère and Kelvin are usually written with small letters, and in the former case abbreviated to amp. In addition to these basic units there are two supplementary units, the radian (rad), for measuring plane angles and the steradian (sr) for solid angles. All other units are known as derived units and can be expressed in terms of the basic and supplementary units. An example is the unit of electric potential, the volt, V, which in basic units becomes $\text{kgm}^2\text{s}^{-3}\text{A}^{-1}$. There are a number of books published about S.I. units with conversions of derived units to basic units included.

In addition to the units themselves there are prefixes indicating multiples or submultiples. An example of such a prefix occurs in the basic unit of the kilogram in which kilo means one thousand. The multiples and submultiples are

Multiple/submultiple	Prefix	Symbol
10^6	mega	M
10^3	kilo	k
10^2	hecto	h
10	deca	da
10^{-1}	deci	d
10^{-2}	centi	c
10^{-3}	milli	m
10^{-6}	micro	μ

Since these are non-dimensional they can be omitted if a dimensional analysis is being undertaken. Whether a length is measured in metres or millimetres makes no difference to the fact that it is a length and as far as dimensions are concerned the symbol m can be used.

Example. The rate of flow of blood (litres per second) through an artery of internal radius r (cm) and length L (m) is given by

$$V = \tfrac{1}{8}\pi \frac{Pr^4}{\eta L}$$

where P $(\mathrm{Nm^{-2}})$ is the pressure difference across the length L and η $(\mathrm{kg\,m^{-1}\,s^{-1}})$ is the dynamic viscosity of blood. Check this equation dimensionally.

The rate of flow V is a volume per unit time and in basic units this will have dimensions $\mathrm{m^3\,s^{-1}}$. On the right-hand side $\tfrac{1}{8}$ and π are non-dimensional constants. The remaining terms give

$$\frac{Pr^4}{\eta L} : \frac{(\mathrm{Nm^{-2}})(m)^4}{(\mathrm{kg\,m^{-1}\,s^{-1}})(m)} = \frac{(\mathrm{kg\,m\,s^{-2}\,m^{-2}})(m)^4}{(\mathrm{kg\,m^{-1}\,s^{-1}})(m)} = \mathrm{m^3\,s^{-1}}$$

and so the equation balances dimensionally.

Dimensional analysis can be used in the original formulation of equations rather than in just checking. This is best illustrated by an example using the physical situation described above.

Example. The rate of flow of blood through an artery is thought to depend upon the radius and length of the artery, the differential pressure, and the viscosity of blood. Use dimensional analysis to investigate possible formulae.

Assume that the flow is proportional to the product of the given parameters each raised to unknown powers.

Therefore, $$V = \alpha r^w L^x P^y \eta^z$$

where α is a non-dimensional constant. If dimensions are now inserted,

$$\mathrm{m^3\,s^{-1}} = \mathrm{m^w\,m^x(kg\,m\,s^{-2}\,m^{-2})^y(kg\,m^{-1}\,s^{-1})^z}$$

$$= \mathrm{m^{w+x-y-z}\,kg^{y+z}\,s^{-2y-z}}.$$

Equating the powers on each side,·

$$3 = w + x - y - z,$$

$$0 = y + z,$$

$$-1 = -2y - z.$$

There are four unknown powers and only three equations and so a complete solution cannot be expected. However from the second equation substituted into the third we have $y = 1$ and so $z = -1$. From the first we are left with $w + x = 3$. Up to this point the formula is

$$V = \alpha r^x L^{3-x} \frac{P}{\eta}.$$

Dimensional analysis can help no further and even so the original form of a product of parameters was an assumption. However, if this assumption was correct, and it seems not unreasonable, we might guess that the flow would be inversely proportional to the length (power of L is minus one) and obtain

$$V = \alpha \frac{P r^4}{\eta L}$$

since $3 - x = -1$ and so $x = 4$. An experiment would yield the value of the non-dimensional constant α and further experiments could confirm or reject this formulation as a whole.

There are some further points which can help with an analysis. Only like quantities can be added together. Metres cannot be added to seconds. This obviously applies to subtraction as well. Exponentials, logarithms, and trigonometric functions are pure numbers and so have no dimensions. Expressions inside these functions must therefore also be dimensionless. Consider for example the population growth relationship

$$P = P_0 2^{2t}$$

developed in §2.6. Both P and P_0 are population numbers and have no dimensions. The exponential 2^{2t} must therefore also be dimensionless as must the exponent itself. The exponent $2t$ can be written as $t/\frac{1}{2}$ where t is elapsed time and the $\frac{1}{2}$ represents the time taken for the population to double. Their quotient is therefore dimensionless. In this form the dimensionless nature of the exponent is clear but in the form $2t$ it was somewhat concealed.

Exercises

1. The periodic time T of a simple pendulum is believed to depend upon the mass M of the bob, the length l of the string, and the local

gravitational acceleration g. Assuming the dependence takes the form $T = kM^x l^y g^z$ where k is a non-dimensional constant, find x, y, and z.

2. A small wading bird, the stilt, weighs about 120 g and has legs approximately 20 cm long. From these two facts it was deduced that another wader, the flamingo, weighing about 1.8 kg, should have legs 3 m long. How was this deduction in error and what should have been the conclusion reached if a correct dimensional argument had been used?

3. In studies of body-heat loss, the surface area of warm-blooded mammals is important. Such areas are difficult to measure and so empirical laws for various species have been formulated to estimate them. Often these laws use as their input data a typical linear measurement, such as body length or height, and the mass M of the mammal concerned. It is assumed that the mass is directly proportional to the volume of the mammal. The area A is then written in the form

$$A = kM^x l^y$$

where k is a constant with dimensions appropriate for converting the mass M to an equivalent volume. Find the relationship between x and y for the above equation to be dimensionally correct.

4. The following mass, length, and surface area data have been collected for a particular species.

Mass M (kg)	2.82	2.28	6.51	5.82	6.17
Length l (cm)	21.9	14.0	31.6	24.5	22.7
Surface area A (cm^2)	1140	960	2020	1810	1870

An empirical law of the form $A = kM^x l^y$ is being sought (cf. question 3). Since k is a constant, the expression

$$k = \frac{A}{M^x l^y}$$

should produce good agreement at all five data points if x (and hence y) is chosen correctly. By trial and error find values of x and y which produce the same value of k, to within 1 per cent, at all five data points. Check the empirical relationship against the extra data point, $A = 1170$ cm^2 when $M = 3.07$ kg and $l = 17.4$ cm, and estimate the surface area if $M = 4.50$ kg and $l = 25.0$ cm.

3.3. Choice of readings

When conducting an experiment the choice of particular values for the independent variable is in the hands of the observer. Usually values within a certain range are to be investigated and it is quite natural to take a set of evenly-spaced values within this range. Depending upon how the results of the investigation are to be treated this may not be the best strategy.

Consider, for example, some experiments to confirm that the photosynthetic rate y of a leaf is related to the intensity of light x falling upon it by an equation of the form

$$y = \frac{1}{a + b/x}$$

where a and b are unknown constants. The light intensity is under the control of the observer and intensities in the range 0 to 2 are to be investigated. Experiments might be conducted in which the intensity of light is increased in steps of 0.200 giving the following table of results.

x	0.000	0.200	0.400	0.600	0.800	1.000	1.200	1.400	1.600	1.800	2.000
y	0.000	0.136	0.176	0.190	0.202	0.205	0.211	0.216	0.217	0.218	0.221

A graph of these results (Figure 3.1) shows that they are tending to

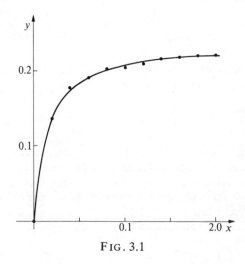

FIG. 3.1

level out and that there are insufficient points in the low light intensity region where the photosynthetic rate is changing most rapidly. The shape of the expected curve can be seen more easily by writing the equation in the form

$$y = \frac{x}{ax+b} = \frac{x/a}{x+b/a} = \frac{1}{a} - \frac{b/a^2}{x+b/a}.$$

Hence

$$\left(x + \frac{b}{a}\right)\left(y - \frac{1}{a}\right) = -\frac{b}{a^2}$$

which is a rectangular hyperbola with a vertical asymptote of $x = -b/a$ and a horizontal asymptote of $y = 1/a$. A sketch is shown in Figure 3.2 in which it is assumed that both a and b are positive. Without knowing

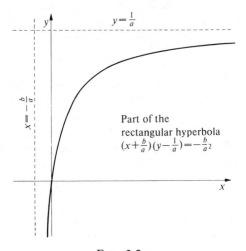

$$y = \frac{1}{a}$$

Part of the
rectangular hyperbola
$$\left(x + \frac{b}{a}\right)\left(y - \frac{1}{a}\right) = -\frac{b}{a^2}$$

$$x = -\frac{b}{a}$$

F IG. 3.2

a and b a particular curve cannot be drawn and so the amounts by which the experimental points deviate from the expected cannot be determined. It would be much more convenient if the expected results lay along a straight line and then if the experimental results also lay along a straight line there would be quite good evidence that theory and experiment were in agreement. In this particular case it is possible to rewrite the equation and transform the variables in such a way that the theoretical results lie on a straight line. Turning the original equation upside-down we have

$$\frac{1}{y} = a + \frac{b}{x}.$$

If the transformations $1/x = u$ and $1/y = v$ are applied the equation becomes

$$v = a + bu$$

which is the equation of a straight line with v-intercept of a and slope b. Not only have we obtained a straight line for the theoretical results but if the experimental results lie on such a line the intercept and slope enable estimates of the unknown constants a and b to be made. A plot of v against u ($1/y$ against $1/x$) should therefore be made.

x	0.000	0.200	0.400	0.600	0.800	1.000	1.200	1.400	1.600	1.800	2.000
u	∞	5.000	2.500	1.667	1.250	1.000	0.833	0.714	0.625	0.555	0.500
y	0.000	0.136	0.176	0.190	0.202	0.205	0.211	0.216	0.217	0.218	0.221
v	∞	7.353	5.682	5.263	4.950	4.878	4.739	4.630	4.608	4.587	4.525

The graph is shown in Figure 3.3. Note that if an intercept on the vertical axis is required the horizontal axis should start at zero otherwise the vertical axis will not be in the correct position and an incorrect reading will result. The points are badly clustered in the bottom left-hand corner and to obtain the best straight line a more even distribution

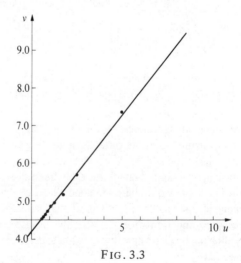

FIG. 3.3

would be desirable. Even so the points do appear to be along a line after allowing for some experimental error.

A more even distribution of u-values would mean a less even distribution of x-values in the original experiment. This would have to be taken into account when it was designed. The range of required x-values (light intensity) ran from 0 to 2 and the corresponding u-values from ∞ to 1.5. Clearly such an infinite range cannot be accommodated so we consider x-values from 0.1 to 2 giving a range of u-values from 10 to 0.5. For more even spacing of u we might consider a set of values such as

u	0.5	1	2	3	4	5	6	7	8	9	10
x	2	1	0.5	0.333	0.250	0.200	0.167	0.143	0.125	0.111	0.100

The x values can be rounded to convenient figures and the gap between $x = 2$ and $x = 1$ filled by including $x = 1.5$. The experiment can be performed with these values to give a set of results which might look like

x	0.100	0.110	0.120	0.140	0.170	0.200	0.250	0.350	0.500	1.000	1.500	2.000
u	10.000	9.091	8.333	7.143	5.882	5.000	4.000	2.857	2.000	1.000	0.667	0.500
y	0.095	0.103	0.109	0.115	0.130	0.136	0.147	0.169	0.186	0.205	0.219	0.221
v	10.523	9.709	9.174	8.696	7.692	7.353	6.803	5.917	5.376	4.878	4.566	4.525

The graphs of y against x and of v against u ($1/y$ against $1/x$) are shown in Figure 3.4 (a) and (b). The spacing in the second graph is now more even at the expense of that in the first. To draw the best straight line through the points on the second graph it is worth remembering that such a line must pass through $(\bar{u}, \bar{v})$ where $\bar{u}$ and $\bar{v}$ are the means of the u and v values respectively. This point is (4.706, 7.101). The line shown crosses the vertical axis at 4.202 and this is therefore the required value of a. The slope can be calculated using two well-spaced points to avoid the loss of significant figures in the subtraction. Hence the slope b is given by

$$b = \frac{9.730 - 4.500}{9.000 - 0.500} = 0.615.$$

The relationship obeyed by the experimental values is therefore

$$y = \frac{1}{4.202 + 0.615/x}.$$

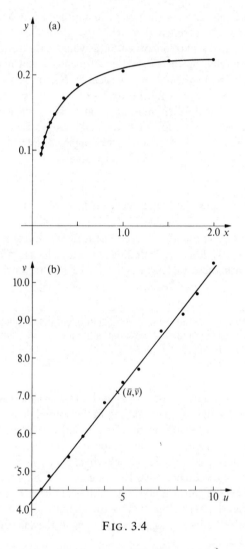

FIG. 3.4

This example has shown that the choice of experimental readings can depend upon the way in which they are to be treated. It has also demonstrated the value of a straight-line graph if this can be obtained by a suitable transformation. To spot such transformations requires some experience and manipulative ability. Each problem has to be considered on its merits and there is no guarantee that an appropriate transformation exists. However, if one is found, it may be possible to use it to determine some unknown constants in the relationship. There are certain standard relationships for which a logarithmic tranformation is used to produce a linear form. Two of these are considered in the following sections.

The straight line in the example has been drawn in such a way that it passes as near as possible to the experimental points, as judged by the eye, while ensuring that it does pass through the mean point $(\bar{u}, \bar{v})$. There is a method of calculating the slope and intercept of such a line without having to draw a graph, although a graph is always useful. This immediately gives the equation of the best straight line. The method is known as the *method of least squares*. It ensures that the sum of the squares of the distances of all the points from the best line is kept to a minimum. The proof is beyond the scope of this book but an outline of the method itself is given in Appendix A.6.

3.4. The log/log graph

If some experimental results have been obtained and it is suspected that the relationship between the two variables might be a *power law* of the form

$$y = Ax^{\alpha},$$

where A and α are constants, this can be tested by a *log/log graph*. Furthermore the constants A and α can be determined. The technique involves taking the logarithms (to any base) of the whole equation to obtain

$$\log y = \log(Ax^{\alpha})$$

$$= \alpha \log x + \log A.$$

The transformations $\log x = u$ and $\log y = v$ convert this to

$$u = \alpha u + \log A$$

which is a linear relationship between u and v ($\log x$ and $\log y$). A graph of $\log y$ against $\log x$ will therefore produce a straight line if a power law relationship does exist between the variables, its slope giving the power α and its intercept on the vertical ($\log y$) axis giving $\log A$ (and hence A).

Example. Groups of fruit fly were kept at different densities (flies per cm^2) and the mean number of eggs produced per day per female were recorded. The results were found to be as follows.

Density D	5	10	20	40	80
No. of eggs N	28.7	22.6	19.5	17.2	14.2

It is believed that a relationship of the form

$$N = AD^\alpha$$

may apply. Test this hypothesis and if it is true determine the constants A and α.

Taking logarithms,

$$\log N = \alpha \log D + \log A.$$

A plot of $\log N$ against $\log D$ will be a straight line if the hypothesis is true.

D:	5	10	20	40	80
$\log D$:	0.6990	1.0000	1.3010	1.6021	1.9031
N:	28.7	22.6	19.5	17.2	14.2
$\log N$:	1.4579	1.3541	1.2900	1.2355	1.1523

A graph of $\log N$ against $\log D$ is shown in Figure 3.5. Note that the graph is drawn with the horizontal axis starting from zero. This is to

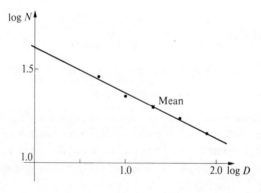

F IG. 3.5

ensure that the vertical axis is in the correct position. If this were not the case an incorrect intercept would result. The points form quite a good line and so the hypothesis is likely to be correct. As in the previous section the means of log D and of log N should be calculated since the best straight line should pass through them. The means give the point (1.301, 1.298) and the intercept of the line with the vertical axis is at log $A = 1.613$ and so $A = 41.02$. Any pair of well-spaced points can be used for the slope and the points chosen here are (0, 1.613) and (1.500, 1.250). The slope is then given by

$$\alpha = \frac{1.250 - 1.613}{1.500 - 0.000} = -0.242.$$

The relationship is therefore

$$N = 41.02D^{-0.242}.$$

It is worth remarking that the densities chosen were not evenly spaced. Each new density is twice the previous one which gives even spacing on a logarithmic scale of log $2 = 0.30103$. If even spacing on a logarithmic scale is required for the independent variable then the values chosen should be in geometric progression (cf. Appendix A.4). Each new point is calculated from its predecessor by multiplying the predecessor by a constant factor, n say. The spacing on a logarithmic scale will then be log n.

Exercises

1. In an electrode system the relationship between current I and applied voltage V is believed to be of the form $I = AV^{\alpha}$ where A and α are constants. Some experiments yield the following results:

V (V)	2	4	8	16	32
I (mA)	4.20	7.07	11.90	20.00	33.75

Test the assumption graphically and if it is correct find appropriate values for A and α.

2. It is thought that the flight-muscle weight of birds is proportional to some power of their body weight. Use the following data to test this hypothesis and if it is supported find the constant of proportionality and the power required.

Body weight (g)	3.8	11.0	17.8	30.0	36.2	79.5	214	260
Flight muscle (g)	0.50	1.95	2.39	5.30	7.65	16.54	29.0	51.7

3. The following table gives the metabolic rates R for animals of various masses M.

M (kg)	0.7	2.0	25	70	120	600	4000
R (kJ/day)	485	695	5570	9630	15 800	55 300	218 000

It is suspected that the rate R is porportional to some power of the mass M. Test this hypothesis and if confirmed find the constants involved.

3.5. The semi-log graph

Exponential functions of one kind or another are fairly common in biology and if such a relationship is suspected it can be checked by the use of a *semi-log graph*. If the form of the relationship is thought to be

$$y = A\, e^{\alpha x},$$

where A and α are constants, taking logarithms to any base gives

$$\log y = \alpha x \log e + \log A.$$

There may be some advantage in using natural logarithms since $\ln e = 1$ and the equation becomes

$$\ln y = \alpha x + \ln A.$$

However, lack of familiarity with natural logarithms can more than off-set this advantage. It will be assumed that common logarithms are being used for this example.

The transformation $\log y = v$ converts

$$\log y = (\alpha \log e)x + \log A$$

into

$$v = (\alpha \log e)x + \log A.$$

This is a linear relationship and a graph of $\log y$ against x, a semi-log graph, will produce a straight line if the original function is an exponential. The slope of the line will be $\alpha \log e$ and its intercept on the vertical axis will be at $\log A$.

Example. A colony of bacteria grow under ideal conditions and the number N of bacteria present at various times t in minutes over a two-hour period is noted. The results show that

t	0	10	25	45	60	75	100	120
N	470	650	1030	1900	3040	4830	10 500	19 500

It is expected that growth is exponential so that

$$N = A^{\alpha t}$$

where A and α are unknown constants. Test this hypothesis and, if it is confirmed, find A and α.

Taking logarithms

$$\log N = (\alpha \log e)t + \log A$$

and so if the hypothesis is true a graph of $\log N$ against t will produce a straight line.

t	0	10	25	45	60	75	100	120
N	470	650	1030	1900	3040	4830	10 500	19 500
$\log N$	2.67	2.81	3.01	3.28	3.48	3.68	4.02	4.29

The graph is shown in Figure 3.6 and the points can be seen to lie on a line. The mean point is (54.4, 3.405) and the line crosses the $\log N$ axis at $\log A = 2.673$ giving $A = 470$. The slope is

$$\alpha \log e = \frac{4\,155 - 2.672}{110 - 0} = 0.01348.$$

Therefore,
$$\alpha = \frac{0.01348}{0.43429} = 0.0310.$$

The relationship is therefore given by

$$N = 470\, e^{0.031t}$$

and the population doubles in a period T where

$$2 = e^{0.031T}.$$

Therefore,
$$\ln 2 = 0.031T.$$

$$T = \frac{\ln 2}{0.031} = 22.4 \text{ minutes.}$$

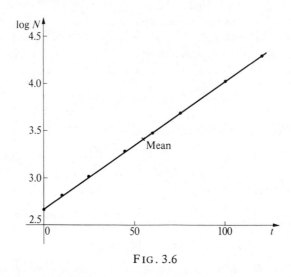

FIG. 3.6

Exercises

1. A micro-organism is believed to grow exponentially with time in its early stages so that the expected relationship between mass m and time t is $m = m_0 e^{\alpha t}$ where m_0 is the initial mass and α a constant. Recordings show

t (h)	0	5	10	15	20	25
m (g)	0.400	0.465	0.540	0.627	0.728	0.848

Confirm the relationship and find m_0 and α. Find the mass m when $t = 50$ assuming the relationship still holds.

2. When light passes through a solution its intensity is reduced. The reduction depends upon the concentration c of the solution and the distance x of liquid through which the light has passed. The relationship is believed to take the form $I = I_0 e^{-\alpha c x}$ where I_0 is the initial intensity. For a solution containing 0.015 moles per litre the following reductions of intensities were found at various distances x.

x (cm)	0	5	10	15	20
I/I_0	1	0.286	0.082	0.023	0.007

Test the expected relationship and if it is found to hold calculate the relative intensity at a penetration of 2.5 cm in the same solution.

3. A laboratory animal of 7 kg mass is given an injection of 25 mg of anaesthetic per kg of body mass. The animal's metabolism clears the anaesthetic from its system over a period. Measurements indicate that the total amount of anaesthetic remaining in the animal over a period of 6 hours are as follows.

Time (h)	0	1	2	3	4	5	6
Anaesthetic (mg)	175	141	120	96	80	67	54

It is suspected that the clearance is exponential. Test this suspicion and if it fits the data, derive an equation for the mass of anaesthetic per kg of body weight remaining at any time *t*.

3.6. Difference tables

Experimental data points usually take the form of a table from which a graph may be plotted. As shown in the previous sections there may be some advantage in transforming the data points in order to compare them with a straight line. If we wish to check how well the data fit a polynomial function, the table of data can be extended using *differences* and conclusions drawn from the form these differences take. Such an extended table, known as a *difference table*, can also be used to detect some kinds of error in the original data.

In order to see how these methods work we need to consider the form that the difference table of a polynomial table takes. As an example consider the cubic

$$f(x) = x^3 + 4x^2 - 3x + 7$$

for values of *x* from 1.0 to 1.5 in steps of 0.1. Such a range can be indicated in a shorter form by writing $x = 1.0(0.1)1.5$, i.e. *x* = first value (step size) last value. For this polynomial we have

x	1.0	1.1	1.2	1.3	1.4	1.5
f(x)	9.000	9.871	10.888	12.057	13.384	14.875

The *first differences* are calculated by subtracting each value of the function from the value to the right of it. Subtracting 9.000 from 9.871 given 0.871, subtracting 9.871 from 10.888 gives 1.017, and so on. *Second·differences* are obtained by a similar subtraction of the first differences. Third and subsequent differences are found in the same way. Rather than stacking these differences vertically it is more convenient to rotate the whole table and write them from left to right.

x	$f(x)$	1st differences	2nd differences	3rd differences
1.0	9.000			
		0.871		
1.1	9.871		146	
		1.017		6
1.2	10.888		152	
		1.169		6
1.3	12.057		158	
		1.327		6
1.4	13.384		164	
		1.491		
1.5	14.875			

The most important point to note is that the third differences are constant. This is a particular case of a general theorem which states that the nth differences of a polynomial of degree n (cf. §2.4) are constant and equal to $n!a_n h^n$ where a_n is the coefficient of x^n and h is the step interval between successive values of the independent variable x. *It is important that there is a constant step interval.* Another point to note is that it is conventional to write only the significant figures of the differences. Preceding zeros and the decimal point are omitted from the table although a difference quoted away from the table should be quoted in full. The last difference is therefore 0.006. From the general theorem this should be equal to $3!a_3 h^3$. For the polynomial under consideration $a_3 = 1$ and $h = 0.1$ so that $3!a_3 h^3 = 0.006$ as expected.

There is some standard terminology applying to difference tables which should be mentioned. It is conventional to denote the initial x value in the table by x_0 and the associated function by f_0. The next x value is denoted by x_1 and the value of the function by f_1 and so on. The *first forward difference* of f_0 is defined to be $f_1 - f_0$ and is denoted by the symbols Δf_0. The first forward difference of f_1 is $f_2 - f_1$ and denoted by Δf_1. The *second forward difference* of f_0 is then defined to be $\Delta f_1 - \Delta f_0$ and denoted by $\Delta^2 f_0$. A difference table of forward differences then looks like

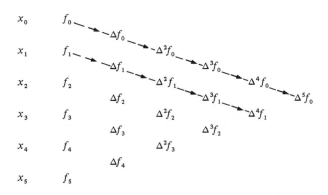

The terminology in no way affects the numbers in a difference table or the way in which they are calculated. It is simply a convenient way of referring to particular differences in a difference table. *Backward differences* and *central differences* can also be defined but will not be used in our discussions.

Returning to the difference table of a polynomial and the general theorem it follows that if a difference table is formed from some data with equal step intervals of the independent variable and the nth differences are constant then the data can be exactly represented by an nth-degree polynomial. Furthermore, if the nth differences are nearly constant then an nth-degree polynomial will give a very close representation. There are reasons however why n should be kept fairly low. One is a matter of simplicity. The lower the degree of the polynomial the simpler it is to find and to use. Another reason is that high-degree polynomials tend to jump up and down quite rapidly and so values calculated at intermediate points may exhibit considerable errors when compared with experimental values at such points.

Example. Test by forming a difference table whether the following data can be represented by a polynomial and if so find the polynomial.

x	0	0.1	0.2	0.3	0.4	0.5
y	0.800	0.705	0.620	0.545	0.480	0.425

The difference table is shown below.

x	y	1st difference	2nd difference
0	0.800		
		−95	
0.1	0.705		10
		−85	
0.2	0.620		10
		−75	
0.3	0.545		10
		−65	
0.4	0.480		10
		−55	
0.5	0.425		

Second differences are constant and so a second-degree polynomial will fit the data. We assume therefore that

$$y = a_2x^2 + a_1x + a_0$$

and wish to find the constants a_2, a_1, and a_0. We can choose any three of the six given points and form three simultaneous differential equations for the constants. In this case however there is a quicker method. The constant difference is 0.010 and so $2!a_2(0.1)^2 = 0.010$. Hence $a_2 = 0.5$. When $x = 0$, $y = 0.800$ and so $a_0 = 0.800$. Using the second point,

$$0.705 = 0.5(0.1)^2 + a_1(0.1) + 0.8.$$

Therefore, $\qquad\qquad\qquad a_1 = -1.$

The polynomial is therefore the quadratic

$$y = 0.5x^2 - x + 0.8.$$

In order to show how a difference table can detect errors in the original data if that data fits or nearly fits a polynomial consider an extended table of the polynomial we first used, $f(x) = x^3 + 4x^2 - 3x + 7$. Assume that the function value at $x = 1.4$ has been recorded as 13.386 instead of 13.384.

x	f(x)	1st diff	2nd diff	3rd diff	4th diff
1.0	9.000				
		0.871			
1.1	9.871		146		
		1.017		6	
1.2	10.888		152		-2
		1.169		8	
1.3	12.057		160		-8
		1.329		0	
1.4	13.386		160		12
		1.489		12	
1.5	14.875		172		-8
		1.661		-4	
1.6	16.536		176		2
		1.837		6	
1.7	18.373		182		
		2.019			
1.8	20.392				

The error has been propagated in a divergent manner so that the fourth differences which should be zero are not and the third differences which should be constant are not either. The errors in the fourth differences are exactly those figures in the table since the differences ought to be zero. The errors in the third differences are the amounts by which the table figures differ from 6

3rd diff errors　　　　0.002　　−0.006　　0.006　　−0.002

4th diff errors　　0.002　　−0.008　　0.012　　−0.008　　0.002

The original error was 0.002 in $f(1.4)$ and dividing through by this shows that the errors have a binomial distribution (Appendix A.5) with alternating sign.

$$1 \quad -3 \quad 3 \quad -1$$

$$1 \quad -4 \quad 6 \quad -4 \quad 1$$

Conversely, if differences occur in one of the columns which are at variance with the general trend, and this variance has a binomial distribution with alternating sign then an error is being propagated. The point of origin of the error can be traced back through the table and its magnitude and sign determined. Errors can be arithmetical in origin and so may propagate from an incorrect difference as well as from a doubtful piece of original data.

Example. Examine the following difference table for errors and correct it if necessary.

x	y	1st diff	2nd diff	3rd diff	4th diff
2.0	0.69315				
		4879			
2.1	0.74194		−227		
		4652		20	
2.2	0.78846		−207		−2
		4445		18	
2.3	0.83291		−189		−3
		4256		15	
2.4	0.87547		−174		−1
		4082		14	
2.5	0.91629		−160		−2
		3922		12	
2.6	0.95551		−148		−1
		3774		11	
2.7	0.99325		−137		−10
		3637		1	
2.8	1.02962		−136		24
		3501		25	
2.9	1.06471		−111		−25
		3390		0	
3.0	1.09861		−111		7
		3279		7	
3.1	1.13140		−104		
		3175			
3.2	1.16315				

The fourth differences seem as if they should all be about −0.00002. If that is the case the last four differences are in error by

$$-0.0008 \qquad 0.0026 \qquad -0.0023 \qquad 0.0009.$$

Taking out a factor of −0.0008 leaves the binomial distribution 1, −3, 3, −1 to a very high degree of accuracy. Tracing the errors back indicates that the first difference 0.03501 is incorrect by −0.0008 and so should have read 0.03509 which can be checked by recalculating it from the data.

Exercises

1. Find in each case, by forming difference tables, the polynomial $P(x)$ which gives rise to the following tables of values.

(a)	x	1	2	3	4	5			
	$P(x)$	−5	−3	−1	1	3			

(b)	x	0	1	2	3	4	5		
	$P(x)$	−3	−6	−5	0	9	22		

(c)	x	0	1	2	3	4	5		
	$P(x)$	2	10	24	44	70	102		

(d)	x	0	1	2	3	4	5	6	7	8
	$P(x)$	0	−2	−4	0	16	50	108	196	320

(e)	x	0	1	2	3	4	5	6	7	8
	$P(x)$	−8	0	0	−2	0	12	40	90	168

2. By forming a difference table find a low degree polynomial which closely fits the following data.

x	1.00	1.05	1.10	1.15	1.20	1.25	1.30	1.35	1.40
y	1.081	0.995	0.907	0.817	0.725	0.631	0.535	0.438	0.340

3. Form a difference table from the following data and hence locate an error in one recording of the dependent variable y.

x	1.80	2.00	2.20	2.40	2.60	2.80	3.00	3.20
y	0.2096	0.2120	0.2141	0.2159	0.2165	0.2187	0.2198	0.2208

x	3.40	3.60	3.80
y	0.2217	0.2225	0.2232

3.7. Interpolation

A situation often arises in which some value of a dependent variable is required at a point which is between values actually tabulated. This may occur in the context of a table of experimental results, in mathematical tables, or in some other table of values. Methods by which an estimate of such an intermediate point can be made are known as *interpolation*. At its simplest level it can mean drawing a smooth graph through experimental or known points and reading off intermediate values from the graph. In many cases this may indeed produce very good results. However, drawing a graph in every situation where an intermediate value is required may not be convenient and for this reason various purely numerical methods of interpolation have been devised.

The simplest assumption that can be made about intermediate points is that they lie on a straight line joining the two end-points. On this assumption the graph illustrating the relationship between two variables would not be a smooth curve, Figure 3.7(a), but would take the form of a series of straight lines linking the points together, Figure 3.7(b). The estimation of an intermediate point in this situation is quite

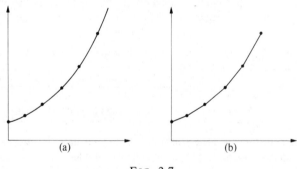

(a)   (b)

FIG. 3.7

simple and can be achieved using similar triangles. Consider Figure 3.8 which shows a pair of end-points joined by a straight line. The first point is (x_0, y_0) and the second (x_1, y_1). We require the value of y at some point x lying between x_0 and x_1 and from the diagram this lies between y_0 and y_1. As in the equation of the straight line (cf. §2.2) the slope of the line can be calculated using both end-points and again using the first end-point and the intermediate point.

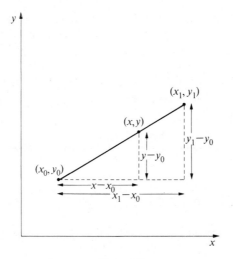

F IG. 3.8

$$\frac{y_1 - y_0}{x_1 - x_0} = \frac{y - y_0}{x - x_0}.$$

Therefore,
$$y = y_0 + \frac{x - x_0}{x_1 - x_0}(y_1 - y_0).$$

The intermediate value can therefore be calculated. This formula can be written in a more compact form using difference notation. The terms $(y_1 - y_0)$ is the first forward difference of y_0 and so can be written Δy_0. The fraction $(x - x_0)/(x_1 - x_0)$ is the proportion of the complete step $x_1 - x_0$ taken up by the intermediate point and so must lie between 0 and 1. If this fraction is denoted by θ the formula can be written

$$y = y_0 + \theta \Delta y_0.$$

If the (unknown) functional relationship between y and x is

$$y = f(x)$$

and the step distance $x_1 - x_0 = h$, the intermediate value of x is $x_0 + \theta h$ and the formula becomes

$$f(x_0 + \theta h) \approx f_0 + \theta \Delta f_0$$

where $y_0 = f(x_0) = f_0$. The only variable is θ and so this is a *linear approximation* and is known as *Newton's forward difference interpolation formula of degree one*.

Example. Estimate $\sqrt[4]{15.6}$ given that $\sqrt[4]{15} = 1.9680$ and $\sqrt[4]{16} = 2.0000$.

x	$\sqrt[4]{x}$	1st difference	
15	1.9680		
		320	$\theta = \dfrac{0.6}{1} = 0.6$
16	2.0000		

Therefore, $\sqrt[4]{15.6} \approx 1.9680 + 0.6 \times 0.0320 = 1.9872.$

The correct answer to four decimal places is 1.9874 and so the interpolation only differs by 2 units in the last decimal place or 0.01 per cent.

An improvement on linear interpolation can be made if a smooth curve can be drawn through the relevant points and this used for interpolation. The simplest curve is the parabola,

$$y = a_0 + a_1 x + a_2 x^2,$$

which requires a knowledge of three points in order to determine the three coefficients a_0, a_1 and a_2. If these three points are (x_0, y_0), (x_1, y_1) and (x_2, y_2), three simultaneous equations can be formed for the coefficients.

$$y_0 = a_0 + a_1 x_0 + a_2 x_0^2,$$
$$y_1 = a_0 + a_1 x_1 + a_2 x_1^2,$$
$$y_2 = a_0 + a_1 x_2 + a_2 x_2^2.$$

Subtracting the first from the second and the second from the third gives

$$a_2(x_1^2 - x_0^2) + a_1(x_1 - x_2) = y_1 - y_0 = \Delta y_0,$$
$$a_2(x_2^2 - x_1^2) + a_1(x_2 - x_1) = y_2 - y_1 = \Delta y_1.$$

The step interval is h and so $x_1 - x_0 = x_2 - x_1 = h$. These two equations can therefore be written

$$a_2 h(x_1 + x_0) + a_1 h = \Delta y_0,$$
$$a_2 h(x_2 + x_1) + a_1 h = \Delta y_1,$$

and subtraction gives

$$a_2 h\{(x_2 + x_1) - (x_1 + x_0)\} = \Delta y_1 - \Delta y_0 = \Delta^2 y_0.$$

Therefore, $$a_2 = \frac{1}{2h^2}\Delta^2 y_0.$$

Hence,

$$a_1 = \frac{1}{h}\Delta y_0 - (x_1 + x_0)\frac{1}{2h^2}\Delta^2 y_0$$

and

$$a_0 = y_0 - x_0\left\{\frac{1}{h}\Delta y_0 - (x_1 + x_0)\frac{1}{2h^2}\Delta^2 y_0\right\} - x_0^2\frac{1}{2h^2}\Delta^2 y_0$$

$$= y_0 - \frac{x_0}{h}\Delta y_0 + \frac{x_1 x_0}{2h^2}\Delta^2 y_0.$$

The quadratic is therefore

$$y = y_0 - \frac{x_0}{h}\Delta y_0 + \frac{x_1 x_0}{2h^2}\Delta^2 y_0 + \frac{x}{h}\Delta y_0 - \frac{x(x_1 + x_0)}{2h^2}\Delta^2 y_0 + \frac{x^2}{2h^2}\Delta^2 y_0$$

$$= y_0 + \frac{x - x_0}{h}\Delta y_0 + \frac{(x - x_1)(x - x_0)}{2h^2}\Delta^2 y_0.$$

If, as before, $x = x_0 + \theta h$ and $y_0 = f(x_0) = f_0$, we have

$$f(x_0 + \theta h) \approx f_0 + \theta \Delta f_0 - \tfrac{1}{2}\theta(1 - \theta)\Delta^2 f_0.$$

This is a *quadratic approximation* and is known as *Newton's forward difference interpolation formula of degree two.*

Example. Estimate $\sqrt[4]{15.6}$ given that $\sqrt[4]{15} = 1.9680$, $\sqrt[4]{16} = 2.0000$, and $\sqrt[4]{17} = 2.0305$.

x	$\sqrt[4]{x}$	1st diff	2nd diff	
15	1.9680			
		320		
16	2.0000		−15	$\theta = \dfrac{0.6}{1} = 0.6$
		305		
17	2.0305			

Therefore,

$$\sqrt[4]{15.6} \approx 1.9680 + 0.6 \times 0.0320 - \tfrac{1}{2} \times 0.6 \times 0.4 \times (-0.0015)$$

$$= 1.9874 \text{ to four decimal places.}$$

In this case the approximation agrees exactly with the correct answer.

It should be noted that the first two terms of this quadratic interpolation formula are the same as the terms in the linear interpolation formula. The additional term in the quadratic formula produces the extra accuracy. Higher degree formulae can be derived in a similar way but they get progressively more complicated. Basically what we are doing is fitting polynomials of progressively higher degree to more and more data points.

Example. A population of insects in a closed environment is studied over a period of seven weeks. Initially there were 100 insects and the number at the end of each week was noted. The following table of results emerged.

Time t (days)	0	7	14	21	28	35	42	49
Population P	100	140	230	390	560	680	740	760

Use Newton's forward difference interpolation formula of degree two to estimate the population sizes at the end of the 17th, 20th, and 40th days.

Two small difference tables are required. The first is for the estimates at $t = 17$ and 20 and the second for $t = 40$.

t	P	1st diff	2nd diff	
14	230			
		160		If $t = 17$, $\theta = \tfrac{3}{7}$.
21	390		10	
		170		If $t = 20$, $\theta = \tfrac{6}{7}$.
28	560			

Hence

$$P(17) \approx 230 + \tfrac{3}{7} \times 160 - \tfrac{1}{2} \times \tfrac{3}{7} \times \tfrac{4}{7} \times 10 = 297 \approx 300$$

and

$$P(20) \approx 230 + \tfrac{6}{7} \times 160 - \tfrac{1}{2} \times \tfrac{6}{7} \times \tfrac{1}{7} \times 10 = 367 \approx 370.$$

t	P	1st diff	2nd diff	
35	680			
		60		
42	740		−40	If $t = 40$, $\theta = \tfrac{5}{7}$.
		20		
49	760			

$$P(40) \approx 680 + \tfrac{5}{7} \times 60 - \tfrac{1}{2} \times \tfrac{5}{7} \times \tfrac{2}{7} \times (-40) = 727 \approx 730.$$

Exercises

1. Given that $\sqrt[5]{45} = 2.14113$, $\sqrt[5]{46} = 2.15056$, and $\sqrt[5]{47} = 2.15983$, use both linear and quadric interpolation to estimate $\sqrt[5]{45.7}$.

2. Referring to the first table of results in §3.3 and Figure 3.1, use linear interpolation, quadratic interpolation, and the graph to estimate the values of y when $x = 0.130, 0.170, 0.280$, and 0.550.

3. Referring to the table of results in §3.5 for bacterial growth use both linear and quadratic interpolation to find the bacteria population after 50 and 57 minutes. Use linear interpolation to estimate the time at which the population reached 4000 and 7500.

3.8. Extrapolation

Whereas interpolation gives a method of estimating intermediate values of a dependent variable using known values of nearby points *extrapolation* attempts to estimate values beyond the range of known values. This is always much more risky because there is no guarantee that the rules or conditions which produced the known points can apply to points outside the range. Rules which apply to dilute solutions may not apply if the solutions become concentrated and so extrapolation into such a region may give completely misleading results. As long as these risks are borne in mind some of the methods of interpolation can be used. A graph can be extended beyond experimental or known values following the same curvature or general shape that is apparent for the known points. Readings can then be taken from the graph. The further from known values that these readings are made the more susceptible they are to serious error.

Interpolation formulae, such as those derived in the previous section, can be used for limited extrapolation. The extrapolation should not be extended beyond one step interval and even within a step interval there may be significant errors.

Example. Given that $\sqrt[4]{15} = 1.9680$ and $\sqrt[4]{16} = 2.0000$ use linear interpolation to estimate $\sqrt[4]{14.5}$ and $\sqrt[4]{16.5}$.

The formula is $f(x_0 + \theta h) \approx f_0 + \theta \Delta f_0$ where in this case $x_0 = 15$, $h = 1, f_0 = 1.9680$, and $\Delta f_0 = 0.0320$. To find $\sqrt[4]{14.5}$ requires $\theta = -0.5$ and to find $\sqrt[4]{16.5}$ requires $\theta = 1.5$. Using the formula gives

$$\sqrt[4]{14.5} \approx 1.9680 + (-0.5) \times 0.0320 = 1.9520,$$

$$\sqrt[4]{16.5} \approx 1.9680 + (1.5) \times 0.0320 = 2.0160.$$

The exact answers, to four decimal places, are 1.9514 and 2.0154 respectively, and so the extrapolated results are both about 0.0006 or 0.03 per cent too high. As a comparison the interpolated result for $\sqrt[4]{15.5}$ is given by

$$\sqrt[4]{15.5} = 1.9680 + 0.5 \times 0.0320 = 1.9840.$$

In this case the exact answer is 1.9842 and so the error is -0.01 per cent or three times better than the extrapolated values.

Example. The variation of fertility y (surviving offspring per 100 mammals) as a function of population density x (mammals per square metre) of small mammals is given by the table

x:	0	1	2	3	4
y:	0	4	12	18	16

Use a linear interpolation formula to find the fertility when $x = 3.5$, 4.5, and 5.0.

If it is assumed that $y = f(x)$ then

$$f(3.5) \approx 18 + 0.5 \times (-2) = 17,$$
$$f(4.5) \approx 18 + 1.5 \times (-2) = 15,$$
$$f(5) \approx 18 + 2 \times (-2) = 14.$$

The tabulated results can be obtained from the formula $y = x^2(5 - x)$ which may be derived using difference table techniques (§3.6). If this formula is used then $f(3.5) = 18.375$, $f(4.5) = 10.125$, and $f(5) = 0$. Clearly, in this case, the extrapolated results are very unreliable.

4 Differentiation

4.1. The slope of a graph

IT IS EASY to define the slope or gradient of a straight line (§2.2) since it is constant. The definition of the slope of a curve is not so obvious because it varies from point to point. This problem can be overcome by associating with each point on a curve a line which runs in the same direction as the curve at that point. The slope of the curve can then be defined to be the same as that of the associated line. The problem is therefore reduced to choosing an appropriate associated line for any specified point on the curve. Such a line is the *tangent* to the curve at the point. There are various ways of describing the tangent to a curve some of which are more precise than others. One way is as the line which just touches but does not cross the curve at the point being considered. This particular definition is not precise enough because it could include lines from any direction which stop at the curve, thereby just touching but not crossing it (Figure 4.1(a)) and exclude genuine tangents such as the x-axis as the tangent to $y = x^3$ at the origin, which does cross the curve (Figure 4.1(b)).

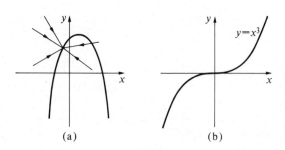

(a)   (b)

FIG. 4.1

A more technically accurate and sophisticated definition is to say that a line is a tangent to a curve at a specified point if it meets the curve at least twice at that point. This requires some explanation which is best accompanied by the use of an example. Any number of lines can be drawn through the point $(1, 3)$ on the parabola $y = 4x - x^2$, most of which will meet the parabola at a second point. Such a line is $y = x + 2$ (Figure 4.2) and the meeting points can be found by solving the equations for the line and the parabola simultaneously.

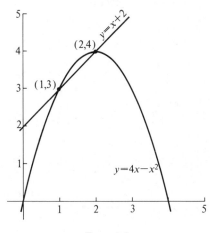

F IG. 4.2

$$y = x + 2,$$
$$y = 4x - x^2.$$

Therefore, $$x^2 - 3x + 2 = 0,$$
$$(x - 1)(x - 2) = 0.$$

The quadratic has two distinct solutions, $x = 1$ and $x = 2$, showing that the line meets the parabola at two distinct point $(1, 3)$ and $(2, 4)$. If the slope of the line is increased, ensuring that the line still passes through the first point, $(1, 3)$, the point at which it meets the parabola again moves down the curve, away from the second point and towards $(1, 3)$. The line $y = \frac{3}{2}x + \frac{3}{2}$ is steeper than $y = x + 2$ and still passes through $(1, 3)$. Solving with the equation of the parabola again gives the second point of meeting (Figure 4.3).

$$y = \tfrac{3}{2}x + \tfrac{3}{2}$$
$$y = 4x - x^2.$$

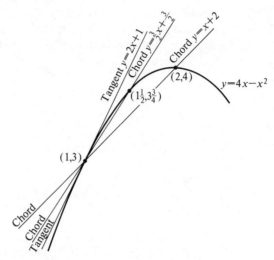

Tangent $y=2x+1$

Chord $y=\frac{3}{2}x+\frac{3}{2}$

Chord $y=x+2$

$(2,4)$

$y=4x-x^2$

$(1\frac{1}{2},3\frac{3}{4})$

$(1,3)$

Chord

Chord

Tangent

F IG . 4.3

Therefore,
$$x^2 - \tfrac{5}{2}x + \tfrac{3}{2} = 0,$$

$$(x-1)(x-\tfrac{3}{2}) = 0.$$

There are again two distinct solutions showing that the line meets the parabola at $(1, 3)$ and $(1\frac{1}{2}, 3\frac{3}{4})$. Now consider the line $y = 2x + 1$ which is steeper still but again passes through $(1, 3)$. The second point of meeting is again found by solving the line with the parabola.

$$y = 2x + 1,$$

$$y = 4x - x^2.$$

Therefore,
$$x^2 - 2x + 1 = 0.$$

$$(x-1)(x-1) = 0.$$

In this case the second solution is the same as the first showing that the second point at which the line meets the parabola is identical with the first. *The line meets the parabola twice at the same point.* This is shown in Figure 4.3. The lines $y = x + 2$ and $y = \frac{3}{2}x + \frac{3}{2}$ are *chords* passing through $(1, 3)$ and meeting the parabola again at a second distinct point. The line $y = 2x + 1$, however, is a *tangent* meeting the parabola twice at the single point $(1, 3)$. The slope or gradient of the parabola

$y = 4x - x^2$ at $(1, 3)$ is the same as that of the tangent line to the parabola at this point, $y = 2x + 1$, and is therefore 2.

Having defined the slope of a graph in terms of its tangents we should perhaps enquire as to whether it has any significance or interpretation and if so whether it can be readily calculated. The interpretation of the slope can best be seen by considering a simple example from mechanics. Neglecting air resistance the distance x fallen by a stone and the time t after it was dropped are related by

$$x = \tfrac{1}{2}gt^2$$

where g is the gravitational acceleration. In S.I. units (§3.2) $g = 9.812$ m s^{-2} but to simplify calculation it will be assumed that $g = 10$ m s^{-2} so that

$$x = 5t^2$$

if x is in metres and t in seconds. A graph of x against t (Figure 4.4) can be drawn from the following table of values.

t	0	1	2	3	4	5
x	0	5	20	45	80	125

After 2 seconds the stone has fallen 20 m and after 4 seconds a total of 80 m. The *average speed* $\bar{v}$ over this period from 2 to 4 seconds after dropping the stone is given by

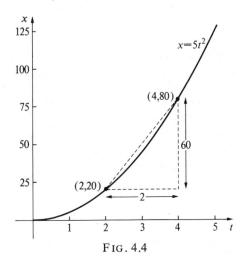

FIG. 4.4

$$\bar{v} = \frac{80 - 20}{4 - 2} = 30 \text{ m s}^{-1}.$$

Geometrically this is the slope of the chord on the graph from $(2, 20)$ to $(4, 80)$. The vertical interval is in metres and the horizontal interval is in seconds and so the slope represents a *rate of change* of metres per second.

If the interval is now reduced to the period from 2 seconds to 3 seconds the average speed is

$$\bar{v} = \frac{45 - 20}{3 - 2} = 25 \text{ m s}^{-1}.$$

The far point has moved down the curve towards the point at 2 seconds. At $2\frac{1}{2}$ seconds $x = 5 \times (2.5)^2 = 31.25$ and the average is now given by

$$\bar{v} = \frac{31.25 - 20}{2\frac{1}{2} - 2} = 22.5 \text{ m s}^{-1}$$

which is represented by the slope of the chord between $(2, 20)$ and $(2.5, 31.25)$. Over the period 2 to $2\frac{1}{4}$ seconds

$$\bar{v} = \frac{25.3125 - 20}{2.25 - 2} = 21.25 \text{ m s}^{-1}$$

and for the period 2 to 2.1 seconds

$$\bar{v} = \frac{22.05 - 20}{2.1 - 2} = 20.5 \text{ m s}^{-1}.$$

In each case as the interval is reduced the second point moves down the curve coming nearer and nearer to the fixed first point at $(2, 20)$. At each stage the slope of the chord, still representing an average speed, becomes more like the slope of the tangent to the curve at $(2, 20)$. If in some systematic way the time interval over which the average speed is being calculated could be reduced to zero, the chord would become a tangent and its slope would be not an average speed but the *instantaneous speed* at $t = 2$ seconds. The two end-points of the interval would have met and the slope of the tangent produced would be the slope of the curve at this point.

4.2. The limit process

Each of the above cases can be included in a more general approach using an arbitrary time point t and an unspecified time interval δt. The Greek letter δ (delta) is conventionally used to represent a change in t, often indicating that the change is small and tending eventually to zero. (Note the use of capital delta, Δ, in difference tables §3.6.) The time interval in this context is from t to $t + \delta t$. The two distances concerned are $5t^2$ and $5(t + \delta t)^2$ so that the average speed is given by

$$\bar{v} = \frac{5(t + \delta t)^2 - 5t^2}{(t + \delta t) - t}$$

$$= \frac{5t^2 + 10t\delta t + 5(\delta t)^2 - 5t^2}{\delta t}$$

$$= 10t + 5\delta t.$$

Note that by putting $t = 2$ and $\delta t = 0.5$ we obtain $\bar{v} = 22.5$ m s^{-1} as in the third case considered above.

Since δt is unspecified we can investigate what happens to the average speed $\bar{v}$ as δt gets smaller and smaller. It is clear from the equation $\bar{v} = 10t + 5\delta t$ that $\bar{v}$ tends to $10t$ as δt tends to zero and that in doing so it ceases to be an average speed and becomes the instantaneous speed v at time t. This is the *limit process* and is indicated symbolically by

$$\underset{\delta t \to 0}{\text{Lim }} \bar{v} = \underset{\delta t \to 0}{\text{Lim }} (10t + 5\delta t) = 10t = v.$$

The instantaneous speed v is the slope of the tangent to the curve, and hence of the curve, at any time t. We have shown that the slope of the distance against time curve, $x = 5t^2$, is given by $v = 10t$ and the method used shows that the slope of any distance against time curve gives the instantaneous speed.

4.3. The first derivative

The approach employed in the previous two sections using distance, time, average speed, and instantaneous speed was considered appropriate because speed is the rate of change with which people are most familiar. It is the rate of change of distance with respect to time; the word 'rate' tends to imply 'with respect to time' as in 'the rate of growth of a population depends upon its environment'. Although many rates of change are with respect to time this is not by any means always the case. We might be interested in the effectiveness of an insecticide on aphids. Graphs of percentage mortality at different concentrations of the insecticide can be drawn and then the rate of change of mortality with respect to concentration could be important. Such a graph is shown in Figure 4.5 where it can be seen that at low concentrations the

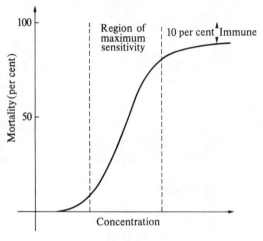

FIG. 4.5

insecticide is almost ineffective. In the central region there is a rapid rate of increase in mortality with respect to concentration and beyond this the rate slows down and the graph levels out. In this situation the slope is important. The region in which it is steepest is the region of maximum sensitivity to the insecticide. Below this region, that is at lower concentrations, the insecticide is wasted because it is ineffective and above this region, at higher concentrations, insecticide is also wasted because lower concentrations would be almost as effective. Either from the empirical data or from some theory describing the action of the insecticide an equation which fits the graph might be found. If this were the case then being able to calculate the slope at any point would be of considerable advantage.

The techniques used to find the slope (speed) of the distance against time graphs can be applied generally. We can consider for instance the graph of

$$y = x^3$$

and examine in detail the interval from x to $x + \delta x$. Over this interval the dependent variable will change from y to $y + \delta y$, say, as shown in Figure 4.6. The slope of the chord from P to Q is $\delta y / \delta x$ and represents the average rate of change of y with respect to x over the interval x

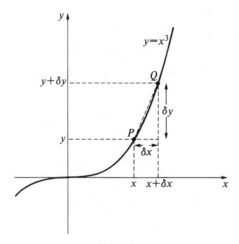

FIG. 4.6

to $x + \delta x$. As δx is progressively reduced the point Q moves down the curve towards P and the slope of the chord PQ becomes more like that of the tangent to the curve at P. If the limiting process can be carried through and

$$\underset{\delta x \to 0}{\text{Lim}} \frac{\delta y}{\delta x}$$

calculated the point Q has met the point P and the slope of the curve at P has been found. Symbolically this slope is denoted by

$$\frac{dy}{dx}$$

so that

$$\frac{dy}{dx} = \lim_{\delta x \to 0} \frac{\delta y}{\delta x}.$$

The slope, dy/dx, of the curve at P represents the (instantaneous) rate of change of y with respect to x and is known as the *first derivative* of y with respect to x.

Applying this to $y = x^3$ the coordinates of P are (x, y) and of Q are $(x + \delta x, y + \delta y)$ where

$$y = x^3$$

and

$$y + \delta y = (x + \delta x)^3.$$

By subtraction of the first equation from the second

$$\delta y = (x + \delta x)^3 - x^3$$

and so

$$\frac{\delta y}{\delta x} = \frac{(x + \delta x)^3 - x^3}{\delta x}$$

$$= \frac{x^3 - 3x^2 \delta x + 3x(\delta x)^2 + (\delta x)^3 - x^3}{\delta x}$$

$$= 3x^2 + 3x\delta x + (\delta x)^2.$$

As $\delta x \to 0$ it is clear that $(\delta y / \delta x) \to 3x^2$ and so

$$\frac{dy}{dx} = \lim_{\delta x \to 0} \frac{\delta y}{\delta x} = 3x^2.$$

There is nothing specific to the examples already considered about the limit process employed. It can be applied to any function and in the most general case of

$$y = f(x)$$

the coordinates of P will still be (x, y) and of Q will still be $(x + \delta x, y + \delta y)$ but the equations relating these variables become

$$y = f(x)$$

and

$$y + \delta y = f(x + \delta x).$$

Hence,

$$\delta y = f(x + \delta x) - f(x)$$

and

$$\frac{\delta y}{\delta x} = \frac{f(x + \delta x) - f(x)}{\delta x}.$$

The *first derivative* of y with respect to x is defined by

$$\frac{dy}{dx} = \lim_{\delta x \to 0} \frac{\delta y}{\delta x} = \lim_{\delta x \to 0} \frac{f(x + \delta x) - f(x)}{\delta x}$$

provided the limit can be evaluated.

Example. If $y = 4x + 7$ find dy/dx by a direct application of the definition of the first derivative.

$$y = f(x) = 4x + 7,$$

$$y + \delta y = f(x + \delta x) = 4(x + \delta x) + 7.$$

Therefore, $\dfrac{dy}{dx} = \lim_{\delta x \to 0} \dfrac{\delta y}{\delta x} = \lim_{\delta x \to 0} \dfrac{4(x + \delta x) + 7 - (4x + 7)}{\delta x}$

$$= \lim_{\delta x \to 0} 4$$

$$= 4.$$

There are other notations for the first derivative in common use which in some circumstances may be advantageous since they are more compact. In the general case of

$$y = f(x)$$

the first derivative of y with respect to x may be written as

$$\frac{dy}{dx} \quad \text{or} \quad y' \quad \text{or} \quad f'(x) \quad \text{or} \quad f'.$$

Such notations are indeed used later in this book. Because rates of change with respect to time are so common a special notation is sometimes used to indicate a derivative with respect to time. If a displacement x was a function of time t we have $x = f(t)$ and the first derivative of x with respect to time might be written as $\dot{x}$, a dot being placed over the x.

4.4. Higher order derivatives

We saw in the previous section that the first derivative with respect to x of x^3 was $3x^2$. This first derivative is a function of x and so its first derivative with respect to x can be found. The process is quite straightforward.

$$y = x^3 \quad \text{and} \quad \frac{dy}{dx} = 3x^2.$$

Let $z = dy/dx$ so that $z = 3x^2$. Applying the definition of the first derivative,

$$\frac{dz}{dx} = \lim_{\delta x \to 0} \frac{\delta z}{\delta x} = \lim_{\delta x \to 0} \frac{3(x + \delta x)^2 - 3x^2}{\delta x}$$

$$= \lim_{\delta x \to 0} (6x + 3\delta x)$$

$$= 6x.$$

Hence $dz/dx = 6x$ but $z = dy/dx$ and substituting this gives the symbols

$$\frac{d}{dx}\left(\frac{dy}{dx}\right) = 6x.$$

This is a first derivative of a first derivative. It is called the *second derivative* or the *derivative of order two* of y with respect to x. The symbols used have become contracted to

$$\frac{d^2y}{dx^2}.$$

A possible route to this contraction, starting from $d/dx(dy/dx)$, is

$$\frac{d}{dx}\left(\frac{dy}{dx}\right) \to \left(\frac{d}{dx}\right)\left(\frac{d}{dx}\right) y \to \left(\frac{d}{dx}\right)^2 y \to \frac{d^2y}{(dx)^2} \to \frac{d^2y}{dx^2}.$$

Other notations in common use are y'', $f''(x)$, and f''.

Given that $y = x^3$ we have shown that $dy/dx = 3x^2$ and $d^2y/dx^2 = 6x$. The second derivative is still a function of x and its derivative with respect to x can be found to give a *third derivative* or *derivative of order three*. Such a derivative would be written as

$$\frac{d^3y}{dx^3}, \quad y''', \quad f'''(x), \quad \text{or} \quad f'''.$$

Obviously the process could be repeated to produce fourth, fifth, sixth, and nth derivatives or derivatives of order 4, 5, 6, and n. If the derivative is of a high order a large number of dashes becomes unwieldy and the symbols used are, for the nth derivative,

$$\frac{\mathrm{d}^n y}{\mathrm{d}x^n}, \quad y^{(n)}, \quad f^{(n)}(x) \quad \text{or} \quad f^{(n)}.$$

Example. Find all derivatives of x^3.

Letting $y = x^3$ we have already found that

$$\frac{\mathrm{d}y}{\mathrm{d}x} = 3x^2$$

and

$$\frac{\mathrm{d}^2 y}{\mathrm{d}x^2} = 6x.$$

Hence,

$$\frac{\mathrm{d}^3 y}{\mathrm{d}x^3} = \lim_{\delta x \to 0} \frac{\delta}{\delta x}\left(\frac{\mathrm{d}^2 y}{\mathrm{d}x^2}\right) = \lim_{\delta x \to 0} \frac{6(x + \delta x) - 6x}{\delta x} = 6$$

and so

$$\frac{\mathrm{d}^4 y}{\mathrm{d}x^2} = \lim_{\delta x \to 0} \frac{\delta}{\delta x}\left(\frac{\mathrm{d}^3 y}{\mathrm{d}x^3}\right) = \lim_{\delta x \to 0} \frac{6 - 6}{\delta x} = 0.$$

All subsequent derivatives are zero.

Each of the higher derivatives has a geometrical interpretation in terms of the derivative immediately preceding it. We know that the first derivative, $\mathrm{d}y/\mathrm{d}x$, is the slope of a graph of y against x. If a graph was then drawn of $\mathrm{d}y/\mathrm{d}x$ against x its slope would be given by $\mathrm{d}^2 y/\mathrm{d}x^2$ and similarly a graph of $\mathrm{d}^2 y/\mathrm{d}x^2$ against x would have slope $\mathrm{d}^3 y/\mathrm{d}x^3$. Such a set of graphs are shown in Figure 4.7 for $y = x^3$.

For a physical interpretation of higher derivatives the distance, time, speed relationships can be used. It has been shown that a graph of distance x against time t has a slope of speed $v = \mathrm{d}x/\mathrm{d}t$. The rate of change of speed with respect to time is known as acceleration and so a graph of v (as $\mathrm{d}x/\mathrm{d}t$) against time would have a slope of acceleration a.

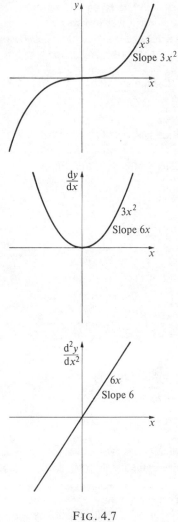

FIG. 4.7

Hence

$$a = \frac{\mathrm{d}v}{\mathrm{d}t}.$$

This makes acceleration the second derivative of distance with respect to time since

$$a = \frac{dv}{dt} = \frac{d}{dt}\left(\frac{dx}{dt}\right) = \frac{d^2x}{dt^2}.$$

Using the dot notation for time derivatives mentioned at the end of the previous section this could be written

$$a = \dot{v} = \ddot{x}.$$

With this analogy it can be seen that a first derivative represents a first rate of change or 'speed'-type term and a second derivative a rate of change of this first rate of change or an "acceleration"-type term.

This branch of mathematics which is concerned with rates of change is known as the *differential calculus*. Since change is so important in all that surrounds us a knowledge of the differential calculus and its complementary branch, the *integral calculus* (Chapter 5), becomes essential for the biological scientist because it provides a precise way of describing change.

4.5. Finding derivatives

Having shown the origins and interpretation of derivatives it remains to be seen whether methods exist for calculating them efficiently. The process of finding or calculating derivatives is known as *differentiation* and in previous sections this has been achieved by a direct application of the definition. This states that if $y = f(x)$ then

$$\frac{dy}{dx} = \lim_{\delta x \to 0} \frac{f(x + \delta x) - f(x)}{\delta x}$$

provided this limit exists. Thus if $y = x^2$,

$$\frac{dy}{dx} = \lim_{\delta x \to 0} \frac{(x + \delta x)^2 - x^2}{\delta x}$$

$$= \lim_{\delta x \to 0} \frac{x^2 + 2x\delta x + (\delta x)^2 - x^2}{\delta x}$$

$$= \lim_{\delta x \to 0} (2x + \delta x)$$

$$= 2x.$$

It has already been shown that, if $y = x^3$, $dy/dx = 3x^2$ and from these two alone it seems reasonable to guess that the derivative of x^n is nx^{n-1}. This is in fact correct and can be proved using the binomial theorem (Appendix A.5).

If $y = x^n$,

$$\frac{dy}{dx} = \lim_{\delta x \to 0} \frac{(x + \delta x)^n - x^n}{\delta x}$$

$$= \lim_{\delta x \to 0} \frac{1}{\delta x} \left(x^n + nx^{n-1} \delta x + \frac{n(n-1)}{2!} x^{n-2} (\delta x)^2 + \ldots - x^n \right)$$

$$= \lim_{\delta x \to 0} \left(nx^{n-1} + \frac{n(n-1)}{2!} x^{n-2} \delta x + \ldots + (\delta x)^{n-1} \right)$$

$$= nx^{n-1}$$

since all terms apart from the first contain factors of δx which cause them to vanish as $\delta x \to 0$. This formula applies also to negative and fractional powers as the following two examples illustrate.

If $y = 1/x$,

$$\frac{dy}{dx} = \lim_{\delta x \to 0} \frac{\dfrac{1}{x + \delta x} - \dfrac{1}{x}}{\delta x} = \lim_{\delta x \to 0} \frac{x - (x + \delta x)}{\delta x(x + \delta x)x}$$

$$= \lim_{\delta x \to 0} \frac{-1}{(x + \delta x)x}$$

$$= -\frac{1}{x^2}$$

$$= (-1)x^{-2} \text{ as the formula would have given.}$$

If $y = \sqrt{x} = x^{\frac{1}{2}}$,

$$\frac{dy}{dx} = \lim_{\delta x \to 0} \frac{\sqrt{(x + \delta x)} - \sqrt{x}}{\delta x}$$

$$= \lim_{\delta x \to 0} \frac{\sqrt{(x + \delta x)} - \sqrt{x}}{\delta x} \times \frac{\sqrt{(x + \delta x)} + \sqrt{x}}{\sqrt{(x + \delta x)} + \sqrt{x}}$$

$$= \lim_{\delta x \to 0} \frac{x + \delta x - x}{\delta x(\sqrt{(x + \delta x)} + \sqrt{x})}$$

$$= \lim_{\delta x \to 0} \frac{1}{\sqrt{(x + \delta x)} + \sqrt{x}}$$

$$= \frac{1}{2\sqrt{x}}$$

$$= \tfrac{1}{2} x^{-\frac{1}{2}} \text{as obtained by the formula.}$$

The establishment of such general formulae as that for differentiating x^{α} where α is any number extends the range of functions which can be differentiated without resorting to the definition of the derivative. Rules exist for differentiating a function multiplied by a constant and for the sum, difference, product, and quotient of two functions. Each of these rules can be proved by a direct application of the definition of the derivative. The rules can be stated as follows.

If $f(x)$ and $g(x)$ are functions of x,

1. The derivative of $kf(x)$, where k is a constant, is $kf'(x)$;
2. The derivative of $f(x) \pm g(x)$ is $f'(x) \pm g'(x)$;
3. The derivative of $f(x)g(x)$ is $f(x)g'(x) + g(x)f'(x)$;
4. The derivative of $f(x)/g(x)$, provided $g(x) \neq 0$, is
$$\frac{g(x)f'(x) - f(x)g'(x)}{(g(x))^2}.$$

Example. Differentiate $4x^3 - 7x^2 + 3x - 2$.

By rule 2 each term can be differentiated separately and by rule 1 the coefficients have no effect. All that remains is to apply the rule for differentiating x^n, and to recall from a previous example (§4.4) that the derivative of a constant is zero. Hence, if $y = 4x^3 - 7x^2 + 3x + 2$,

$$\frac{dy}{dx} = 4 \times 3x^2 - 7 \times 2x + 3 \times 1$$

$$= 12x^2 - 14x + 3\cdot$$

Example. Find the slope of $y = x/(x-1)$ at any point.

Using rule 4 with $f(x) = x$ and $g(x) = x - 1$ gives

$$\frac{dy}{dx} = \frac{(x-1) \times 1 - x \times 1}{(x-1)^2} = \frac{-1}{(x-1)^2}$$

and is valid provided the point $x = 1$ is excluded.

These rules enable us to differentiate any polynomial and rational functions but cannot help with an expression such as

$$\sqrt{(x^2 + 2)}.$$

Here there is one function, a quadratic, buried inside another, a square root. All such functions are included in the general expression

$$f(g(x)).$$

To investigate how the derivatives of such functions can be found we consider the equation

$$y = f(g(x))$$

and let $g(x) = u$ so that $y = f(u)$. From the definition of the derivative,

$$\frac{dy}{dx} = \lim_{\delta x \to 0} \frac{f(g(x + \delta x)) - f(g(x))}{\delta x},$$

but since $g(x) = u$, $g(x + \delta x) = u + \delta u$ and so

$$\frac{dy}{dx} = \lim_{\delta x \to 0} \frac{f(u + \delta u) - f(u)}{\delta x} = \lim_{\delta x \to 0} \frac{f(u + \delta u) - f(u)}{\delta u} \times \frac{\delta u}{\delta x}.$$

Now $f(u + \delta u) - f(u) = \delta y$ and $g(x + \delta x) - g(x) = \delta u$. The second means that as $\delta x \to 0$ so must δu and the first enables us to write

$$\frac{dy}{dx} = \lim_{\delta x \to 0} \frac{\delta y}{\delta u} \times \frac{\delta u}{\delta x}$$

$$= \lim_{\delta u \to 0} \frac{\delta y}{\delta u} \ \lim_{\delta x \to 0} \frac{\delta u}{\delta x}$$

$$= \frac{dy}{du} \cdot \frac{du}{dx}.$$

The formula

$$\frac{dy}{dx} = \frac{dy}{du} \cdot \frac{du}{dx}$$

is known as the *chain rule* and can be extended if there are three functions buried one inside the other to give

$$\frac{dy}{dx} = \frac{dy}{du} \cdot \frac{du}{dv} \cdot \frac{dv}{dx}.$$

We are now in a position to differentiate $\sqrt{(x^2 + 2)}$. We let $y = (x^2 + 2)^{\frac{1}{2}}$ and $u = x^2 + 2$ so that $y = u^{\frac{1}{2}}$. Then

$$\frac{dy}{du} = \tfrac{1}{2} u^{-\frac{1}{2}} \text{ and } \frac{du}{dx} = 2x$$

and so

$$\frac{dy}{dx} = \frac{dy}{du} \cdot \frac{du}{dx} = \tfrac{1}{2} u^{-\frac{1}{2}} \times 2x = \frac{x}{\sqrt{u}}$$

$$= \frac{x}{\sqrt{(x^2 + 2)}}.$$

The chain rule makes it possible to differentiate functions of y with respect to x. The equation $y = x^2$ could be rewritten $x = \sqrt{y}$. Differentiating both sides of the equation with respect to x gives

$$1 = \frac{d}{dx}(\sqrt{y}) = \frac{d}{dy}(\sqrt{y}) \times \frac{dy}{dx}$$

$$= \tfrac{1}{2} y^{-\frac{1}{2}} \frac{dy}{dx} = \frac{1}{2\sqrt{y}} \frac{dy}{dx}$$

and so

$$\frac{dy}{dx} = 2\sqrt{y}.$$

Now $\sqrt{y} = x$ so that this expression for dy/dx is the same as that obtained by differentiating $y = x^2$ directly to give $dy/dx = 2x$. Differentiating a whole equation without first solving for the dependent variable is known as *implicit differentiation*. There may be situations where this is the only way of finding a derivative.

Example. Use implicit differentiation to find dy/dx for the equation of the circle $x^2 + y^2 = a^2$.

Differentiating the whole equation gives

$$2x + 2y\frac{dy}{dx} = 0.$$

Therefore,

$$\frac{dy}{dx} = -\frac{x}{y}.$$

In order to differentiate exponential and logarithmic functions a key result is required which will be quoted here but not explained until §5.4.

$$\text{If } y = \ln x, \, dy/dx = 1/x.$$

This statement itself gives the derivative of a natural logarithm and since exponentials and logarithms are closely related (§2.7) it can be used to find the derivative of an exponential. If $y = e^x$, taking natural logarithms of both sides gives

$$\ln y = x.$$

Differentiating implicitly with respect to x gives

$$\frac{d}{dx}(\ln y) = 1.$$

Therefore,

$$\frac{d}{dy}(\ln y) \times \frac{dy}{dx} = 1.$$

However,

$$\frac{d}{dy}\ln y = \frac{1}{y},$$

and so
$$\frac{1}{y}\frac{dy}{dx} = 1$$

or
$$\frac{dy}{dx} = y = e^x.$$

This means that differentiating e^x leaves it unchanged. The slope of the graph of $y = e^x$ is e^x. If a different base was used the process is a little more complicated. First the base is converted to e and then the procedure above is adopted using the chain rule.

Example. Find the derivative of 2^x.

Let $y = 2^x$. Converting this to base e (§2.7) gives
$$y = e^{x \ln 2}.$$

Now let $u = x \ln 2$ and differentiate using the chain rule.
$$y = e^u,$$
$$\frac{dy}{dx} = \frac{dy}{du} \times \frac{du}{dx}$$
$$= e^u \times \ln 2 = e^{x \ln 2} \ln 2$$
$$= 2^x \ln 2.$$

More generally if $y = a^x$, $dy/dx = a^x \ln a$.

Example. The isotope of Sodium, Na^{24}, has a half life of 15 hours. If initially there was 7.5 g of this isotope then the mass m remaining after a time t hours has elapsed is given by
$$m = 7.5(\tfrac{1}{2})^{t/15}.$$

Find an expression for the rate of change of m as a function of t and the mass and its rate of change when $t = 24$.

We require to find dm/dt. We may write
$$m = 7.5\{(\tfrac{1}{2})^{1/15}\}^t$$

so that
$$\frac{dm}{dt} = 7.5\{(\tfrac{1}{2})^{1/15}\}^t \ln\{(\tfrac{1}{2})^{1/15}\}$$

$$= -\frac{\ln 2}{2}(\tfrac{1}{2})^{t/15}.$$

The sign of this derivative is always negative showing that the mass m is always decreasing.

When $t = 24$,

$$m = 7.5(\tfrac{1}{2})^{24/15} = 2.47g$$

and

$$\frac{dm}{dt} = -\frac{\ln 2}{2}(\tfrac{1}{2})^{24/15} = -0.114gh^{-1}.$$

The chain rule can be used to find the derivative of $e^{f(x)}$. Let $y = e^{f(x)}$ and $f(x) = u$ so that $y = e^u$

$$\frac{dy}{dx} = \frac{dy}{du} \cdot \frac{du}{dx}$$

$$= e^u f'(x) = e^{f(x)} f'(x).$$

Example. Differentiate e^{x^2}.

Using the above formula the derivative is $2x\,e^{x^2}$.

The chain rule can also be used to produce some general formulae for derivatives involving logarithmic functions. Assume we require the derivative of $\ln(f(x))$. Let $y = \ln(f(x))$ and $f(x) = u$ so that $y = \ln u$

$$\frac{dy}{dx} = \frac{dy}{du} \cdot \frac{du}{dx}$$

$$= \frac{1}{u} \cdot f'(x) = \frac{f'(x)}{f(x)}.$$

If we had $\log_a(f(x))$ the base is first converted to e to give a natural logarithm. From §2.7, property 8, we can write

$$\log_a(f(x)) = \frac{\ln(f(x))}{\ln a}.$$

The previous result can now be used to give

$$\frac{dy}{dx} = \frac{1}{\ln a} \cdot \frac{f'(x)}{f(x)} = \log_a e \cdot \frac{f'(x)}{f(x)}.$$

Example. Find the derivative of $\ln(x^3 - 2x + 1)$.

Using the general result that

$$\frac{d}{dx}(\ln f(x)) = \frac{f'(x)}{f(x)}$$

gives

$$\frac{dy}{dx} = \frac{3x^2 - 2}{x^3 - 2x + 1}.$$

Derivatives of $\sin x$ and $\cos x$ can be found directly from the definition of the derivative and derivatives of other trigonometric functions by application of the various rules. If $y = \sin x$ then

$$\frac{dy}{dx} = \lim_{\delta x \to 0} \frac{\sin(x + \delta x) - \sin x}{\delta x}$$

$$= \lim_{\delta x \to 0} \frac{2 \cos\left(x + \dfrac{\delta x}{2}\right) \sin\left(\dfrac{\delta x}{2}\right)}{\delta x}$$

$$= \lim_{\delta x \to 0} \cos\left(x + \frac{\delta x}{2}\right) \lim_{\delta x \to 0} \frac{\sin\left(\dfrac{\delta x}{2}\right)}{\left(\dfrac{\delta x}{2}\right)}$$

$$= \cos x.$$

This proof involves a trigonometric identity for the difference of two sines and a knowledge that the limit of $\sin x/x$ as x tends to zero is 1 provided x is in radians. It can be shown by a similar argument that

$$\frac{d}{dx}(\cos x) = -\sin x.$$

Alternatively, if $\cos x$ is written as $\sqrt{(1 - \sin^2 x)}$ (§2.8),

$$\frac{\mathrm{d}}{\mathrm{d}x}(\cos x) = \frac{\mathrm{d}}{\mathrm{d}x}\{\sqrt{(1-\sin^2 x)}\}$$

$$= \tfrac{1}{2}(1-\sin^2 x)^{-\frac{1}{2}} \times (-2\sin x \cos x)$$

by the chain rule, and simplification gives the previous result. The derivative of $\tan x$ can be found by recalling that $\tan x = \sin x/\cos x$ and using the rule for a quotient (rule 4).

$$\frac{\mathrm{d}}{\mathrm{d}x}(\tan x) = \frac{\mathrm{d}}{\mathrm{d}x}\left(\frac{\sin x}{\cos x}\right)$$

$$= \frac{\cos x \cos x - \sin x(-\sin x)}{\cos^2 x}$$

$$= \frac{1}{\cos^2 x} = \sec^2 x.$$

Other results are given in a table of derivatives in Appendix B.2.

Example. Find the derivative of $\sin(1/(1+x))$.

If $y = \sin(1/(1+x))$ let $u = 1/(1+x)$ so that $y = \sin u$

$$\frac{\mathrm{d}y}{\mathrm{d}x} = \frac{\mathrm{d}y}{\mathrm{d}u}\frac{\mathrm{d}u}{\mathrm{d}x}$$

$$= \cos u \times \frac{-1}{(1+x)^2}$$

$$= -\frac{\sin\left(\dfrac{1}{1+x}\right)}{(1+x)^2}.$$

Example. A population of large mammals experiences minor sinusoidal variations with a one year period and major sinusoidal variations with an eleven year period. The size P of the population can be represented by the formula

$$P = P_0 \left\{ 1 + \frac{1}{5}\sin(2\pi t) + \frac{1}{2}\sin\left(\frac{2}{11}\pi t\right) \right\}$$

where P_0 is the mean size and t the time in years. If P_0 is 5000 find at what rate the population is changing after $2\frac{1}{2}$, $2\frac{3}{4}$, 3, and $7\frac{1}{2}$ years. Find also the size of the population at each of these times.

The rate of change of the population is given by

$$\frac{\mathrm{d}P}{\mathrm{d}t} = P_0 \left\{ \frac{1}{5} \times 2\pi \times \cos(2\pi t) + \frac{1}{2} \times \frac{2\pi}{11} \times \cos\left(\frac{2\pi}{11}t\right) \right\}$$

$$= 2\pi P_0 \left\{ \frac{1}{5}\cos(2\pi t) + \frac{1}{22}\cos\left(\frac{2\pi}{11}t\right) \right\}.$$

The rate will be in units of number per annum. Since fairly specific times are given a more useful unit might be number per day. The rate per day is 1/365th of the rate per year.

Using the formulae for P and $\mathrm{d}P/\mathrm{d}t$ the following table can be calculated

Time t	Population P	Annual rate $\dfrac{\mathrm{d}P}{\mathrm{d}t}$	Daily rate $\dfrac{1}{365}\dfrac{\mathrm{d}P}{\mathrm{d}t}$
$2\frac{1}{2}$	7475	−6080	−16.7
$2\frac{3}{4}$	6500	0	0
3	7475	6080	16.7
$7\frac{1}{2}$	2726	−6876	−18.8

A negative rate of change means that the population is decreasing. Hence when $t = 7\frac{1}{2}$ years the population is 2726 and decreasing at the rate of 18.8 per day.

Exercises

1. If $y = f(x)$, the first derivative $\mathrm{d}y/\mathrm{d}x$ is defined by

$$\frac{\mathrm{d}y}{\mathrm{d}x} = \lim_{\delta x \to 0} \frac{f(x + \delta x) - f(x)}{\delta x}.$$

Use this definition directly to find $\mathrm{d}y/\mathrm{d}x$ if $f(x)$ is given by

(a) $x + 7$; (b) $x^2 + 2$; (c) $(x + 1)^2$;

(d) $x(2x + 1)$;   (e) $x^2 - 4x + 7$;   (f) $\dfrac{1}{x + 1}$:

(g) $\dfrac{1}{x^2}$;   (h) $\sqrt{(x + 4)}$;   (i) $\sqrt{(x^2 + 2x - 3)}$;

(j) $\dfrac{1}{\sqrt{x}}$.

2. Use rules 1 to 4 for differentiation to find the first derivatives with respect to x of the following functions.

(a) $3x$;

(b) $x^2 - 3$;

(c) $x^3 - 4x^2 + 7x - 5$;

(d) $x(x + 1)$;

(e) $(x^2 + 7)(x^3 - 5)$;

(f) $(x^2 - 3x + 7)(x^5 - 2x^2 + 1)$;

(g) $\dfrac{x^2}{x + 3}$;

(h) $\dfrac{x - 3}{x + 4}$;

(i) $\dfrac{x^2 - 3x + 2}{x^2 + 2x + 1}$;

(j) $\dfrac{(x - 1)(x^2 - 6x + 5)}{x^4 - 3x^2 + 5}$.

3. Use the chain rule to find the first derivatives of:

(a) $(2x^2 - 3x + 7)^{10}$;   (b) $\sqrt{(x^3 - 2x + 3)}$;   (c) $(4x + 3)^{\frac{1}{3}}$;

(d) $(x^2 - 5x + 7)^{\frac{4}{5}}$;   (e) $(x^3 - 2x + 1)^{-\frac{2}{3}}$.

4. Use implicit differentiation to find dy/dx if x and y are related by the following equations.

(a) $3x^2 - 2y^2 = 5$;

(b) $x^{\frac{2}{3}} + y^{\frac{2}{3}} = a^{\frac{2}{3}}$;

(c) $x^2 - 3xy + 2y^2 = 8$;

(d) $x^2y^3 - x^3y^2 = 1$;

(e) $x^4 + 2x^2y^2 + y^4 = 4$.

5. Differentiate

(a) e^{2x};

(b) e^{2x+1};

(c) e^{x^2+2x+1};

(d) 2^{2x};

(e) 10^{x^2-3};

(f) $\ln(x + 1)$;

(g) $\ln(x^2 + 2x - 1)$;   (h) $\sin x + \cos x$;   (i) $\sin(2x)$;

(j) $\sin^3 x$; (k) $\sin(x^2)$; (l) $x \sin x$;

(m) $\dfrac{\sin x}{x}$; (n) $\ln(\cos x)$; (o) $\operatorname{cosec} x$;

(p) $x \ln x - x$; (q) $\cos(1 - 3x - x^2)$; (r) $\sqrt{\sin(4x + 1)}$;

(s) $\ln(\sin 2x - x^2)$; (t) $2^{\ln(3x-1)}$

4.6. Maxima and minima

In some predator and prey situations the populations of both are found to be periodic each with the same period but with a phase difference. The phase difference is due to the fact that it takes some time before a change in the prey population is reflected in a similar change in the predator population. An increase in the prey population is found some time afterwards to be followed by an increase in the predator population. This increase cannot be fully supported by the prey population which therefore falls. This worsens the situation for the predator which becomes less fertile or loses a greater percentage of its young. The predator population therefore falls and the natural growth of the prey population is no longer offset by the influence of a large predator population. The prey population increases and the cycle repeats itself. In making a mathematical model or representation of this situation it is essential to be able to find out where the maxima and minima of the two populations occur in order to compare these with the observed situation. If the model is good enough it would be possible to predict not only where the maxima and minima are but also how large or small they become. If this degree of sophistication can be reached then the influence of other factors can be gradually included. One such factor might be the availability of food for the prey population, an availability which is perhaps decreased by man's expanding agriculture in the area. In this situation it could be vital to predict the likely minima reached by both populations to ensure that the critical level below which there is a risk of extinction is not reached.

The above example is one illustration of numerous situations where it can be important to know where maxima and minima occur. This section is devoted to the methods used to locate maxima and minima by application of the differential calculus. Before searching for maxima and minima we ought to be clear about what we are looking for. Figure 4.8 shows the graph of a function which goes up and down. The maximum point could reasonably be taken to be the highest point that

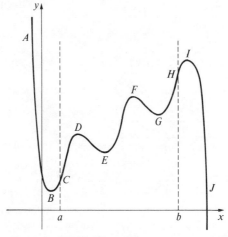

FIG. 4.8

the graph reaches, the point A, which goes upwards to infinity of a vertical asymptote. However, this occurs at a negative x value and we may only be interested in values of x within the range $a \leqslant x \leqslant b$. In this case the peak I would also be excluded. There are in fact a number of peaks (and troughs) each being the highest point in its immediate neighbourhood. The *absolute maximum* within the range might be defined to be the highest of all these various *local maxima*. Such a point would be F. This alone is not sufficient because H, although not a peak, is still higher than F. We must therefore also take into account the value that the function takes at the end-points of the range being considered as well as the values of the function at local maxima. Selecting the largest of all these values will give the absolute maximum within the range $a \leqslant x \leqslant b$. Absolute minima can be defined in a similar way. Note that some local minima may be higher than some local maxima. G is higher than D.

Once the end-points have been decided upon, it is a simple matter to calculate the value that the function takes at these end-points. What is not quite so simple is to decide upon exact mathematical criteria for there to be a maximum or a minimum at some point. In the previous discussion a local maximum was taken to be the highest point in a particular neighbourhood and took the form of a rounded peak. Sharp or spiky peaks will be excluded because nearly all of the functions likely to be encountered will be continuous and reasonably smooth.

Sudden angular changes of direction are not very common. If one was walking a mountain chain which had contours like the curve in Figure 4.8 the approach to a peak would be a climb, the top would be flat, and then a descent is made on the far side. If such a description could be put into mathematical terms it would provide a suitable definition of a local maximum

A peak is shown in Figure 4.9(a). At the very top the curve is running horizontal because it has just stopped rising and is about to

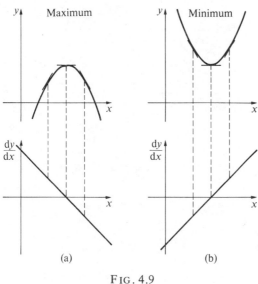

F IG . 4.9

fall. The tangent to the curve at the highest point is horizontal; its slope is zero. We might therefore search for local maxima by looking for any point at which the slope of the curve, that is its first derivative dy/dx, is zero. If we now consider Figure 4.9(b) we can see that the slope is also zero at a local minimum. Searching for points of zero slope will therefore detect local maxima and local minima as well as certain other points of the type shown in Figure 4.10 which are known as *points of inflexion*. Points at which the slope dy/dx is zero are known as *critical points* or *turning points*. The term critical point will be used here because turning point has associated with it a slight implication that the graph turns over at such a point. This is true at local maxima and minima but not so at points of inflexion. Critical points can

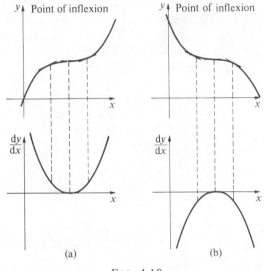

FIG. 4.10

therefore be found by equating the first derivative to zero and solving the resulting equation. Hence if

$$y = f(x),$$

the critical points are the solutions of

$$f'(x) = 0.$$

In order to distinguish maxima from minima and both from points of inflexion a more detailed look at their features is required. Returning to Figure 4.9(a) we see that just before a maximum the graph is rising, its slope is therefore positive, at the maximum it is horizontal, slope zero, and just beyond the graph is falling, its slope is negative. The first derivative is therefore initially positive, becomes zero at the maximum, and then negative. A graph of dy/dx against x therefore falls from left to right, passing through the x-axis at the maximum as shown in the lower part of Figure 4.9(a). The slope of this graph is therefore negative. The slope is however the first derivative of the variable plotted on the vertical axis, namely dy/dx. Its first derivative is d^2y/dx^2 and so at a local maximum

$$f''(x) < 0.$$

The situation at a local minimum is just the reverse. The slope is initially negative, becomes zero, and then positive so at a local minimum (Figure 4.9(b))

$$f''(x) > 0.$$

At points of inflexion the slope is either positive, becoming zero, and then returning positive or negative, becoming zero, and returning negative. The graphs of dy/dx against x look like a minimum in the first case and a maximum in the second (Figure 4.10). In both cases, however, the curves are horizontal at the point of inflexion and so the slope is zero. Since this is a graph of dy/dx against x a zero slope means that at a point of inflexion

$$f''(x) = 0.$$

It should perhaps be mentioned that at certain types of local maxima and minima the second derivative can be zero. While it is certainly true that if $f'(x) = 0$ and $f''(x) < 0$ there is a local maximum and if $f'(x) = 0$ and $f''(x) > 0$ there is a local minimum, these conditions do not cover every possible case. It can be shown that if the second derivative is zero the fourth derivative should be found and its sign examined. If this is still zero the sixth derivative should be examined. In general there is a local maximum if $f'(x) = 0$ and the first non-zero *even* derivative is negative. There is a local minimum if $f'(x) = 0$ and the first non-zero *even* derivative is positive. There is a point of inflexion if all *even* derivatives are zero. Fortunately in the majority of cases these extra complications do not arise.

Example. A rectangular enclosure is to be made using 400 m of wire fencing. What dimensions should it be in order to fence in the maximum area? How is this result affected if a long straight hedge can be used as one boundary?

In the first case the enclosure will take the form shown in Figure 4.11(a) where $0 \leqslant x \leqslant 200$. If the area is A then

$$A = x(200 - x) = 200x - x^2.$$

Critical points are given by $dA/dx = 0$ and so

$$\frac{dA}{dx} = 200 - 2x = 0.$$

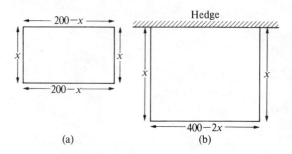

(a)   (b)

FIG. 4.11

Therefore, $x = 100.$

Whether this is a local maximum, minimum, or a point of inflexion may be determined by the second derivative

$$\frac{d^2A}{dx^2} = -2 < 0.$$

Hence the second derivative is negative indicating a local maximum. The only critical point is at $x = 100$ and so this is the local maximum point giving $A = 10\,000$ m². At each of the end-points, $x = 0$ and $x = 200$, the area $A = 0$ and so the local maximum is the absolute maximum.

The situation in the second case is shown in Figure 4.11(b)

$$A = x(400 - 2x) = 2x(200 - x).$$

Note that this is exactly twice the expression for the previous area and so we can expect the final answer to be exactly twice the previous maximum. We will confirm this using differentiation.

$$\frac{dA}{dx} = 400 - 4x = 0.$$

Therefore, $x = 100.$

$$\frac{d^2A}{dx^2} = -4 < 0 \text{ and so a local maximum.}$$

As before the area at each end-point, $x = 0$ and $x = 200$, is zero and so there is an absolute maximum of $A = 20\,000$ m² at $x = 100$.

The dimensions of the enclosure in the first case are 100 m × 100 m giving an area of 10 000 m². In the second case the dimensions are 100 m × 200 m with the hedge being used as one of the longer boundaries. The area is now 20 000 m².

Example. Assuming the thin lens formula

$$\frac{1}{u} + \frac{1}{v} = \frac{1}{f}$$

where u is the object distance, v the image distance, and f the focal length find the minimum separation that can exist between a real object and a real image if the lens is assumed to be convex.

This is the same problem as described in an example in §2.5 but in this case differential calculus will be used to solve it. Since the object and image are both real and the lens is convex u, v, and f are all positive and a typical ray diagram is shown in Figure 4.12. If the separation is

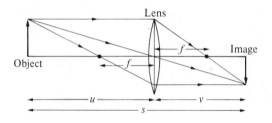

F ig. 4.12

s then $u + v = s$. Hence, in the thin lens formula either u or v can be eliminated and a relationship between s and either v or u formed. The focal length f is a constant. We shall eliminate v.

$$\frac{1}{u} + \frac{1}{s-u} = \frac{1}{f} \text{ and so } \frac{1}{s-u} = \frac{1}{f} - \frac{1}{u} = \frac{u-f}{uf}.$$

Therefore,

$$s = u + \frac{uf}{u-f}.$$

For critical points $ds/du = 0$.

$$\frac{ds}{du} = 1 + \frac{(u-f)f - uf}{(u-f)^2} = 1 - \frac{f^2}{(u-f)^2} = 0.$$

Hence, $$(u - f)^2 = f^2.$$

Therefore, $$u = 0 \text{ or } 2f$$

$$\frac{d^2s}{du^2} = \frac{2f^2}{(u-f)^3} > 0 \text{ if } u = 2f\text{: local minimum,}$$

$$< 0 \text{ if } u = 0 \text{ : local maximum.}$$

For the image to be real $u \geqslant f$. If $u \to f$, $s \to \infty$. At the other extreme if $u \to \infty$, $s \to \infty$. In both situations the separation is a maximum and so the end-points do not provide minima. The local minimum at $u = 2f$ must be the absolute minimum giving a minimum separation

$$s = 4f.$$

Example. The size P of a population after t years in given by (cf. example in §4.5)

$$P = 5000 \left(1 + \frac{1}{5}\sin(2\pi t) + \frac{1}{2}\sin\left(\frac{2\pi}{11}t\right)\right).$$

Find the time at which the first local maximum of this population occurs and its size at this time.

We require to solve $dP/dt = 0$ and select the smallest solution for which $d^2P/dt^2 < 0$.

$$\frac{dP}{dt} = 5000\left(\frac{1}{5} \times 2\pi \cos(2\pi t) + \frac{1}{2} \times \frac{2\pi}{11}\cos\left(\frac{2\pi}{11}t\right)\right).$$

This vanishes when

$$22\cos(2\pi t) + 5\cos\left(\frac{2\pi}{11}t\right) = 0.$$

There is no obvious way of solving this equation. By looking at the original expression for P and noting that $\sin(2\pi t)$ has its first maximum at $t = \frac{1}{4}$ we might expect the first local maximum of P to be near this value. We could try values such as $t = 0.26, 0.27, 0.28$ and so on to obtain a better estimate. However, there are numerical techniques for solving such equations, one of which is discussed in §4.10. Further discussion of this problem will be deferred until then.

An ability to locate local maxima and minima can be of considerable value in curve sketching.

Example. Sketch the graph of $y = 1 - 2x - 3x^2 - x^3$.

This is a cubic (§2.4) with a general trend of falling from left to right. Any local maxima and minima can be located using the differential calculus. At critical points

$$\frac{dy}{dx} = -2 - 6x - 3x^2 = 0,$$

$$x = \frac{-6 \pm \sqrt{(36 - 4 \times 3 \times 2)}}{6} = -1 \pm \frac{1}{\sqrt{3}}$$

$$= -1.58 \text{ or } -0.42,$$

$$\frac{d^2y}{dx^2} = -6 - 6x = -6(1 + x).$$

$$\text{At } x = -1.58, \frac{d^2y}{dx^2} > 0: \text{minimum.}$$

$$\text{At } x = -0.42, \frac{d^2y}{dx^2} < 0: \text{maximum.}$$

Hence, there is a local minimum at $(-1.58, 0.62)$, and a local maximum at $(-0.42, 1.38)$. The graph is shown in Figure 2.14 of §2.4.

Exercises

1. Use differential calculus to determine whether the following expressions for y have any local maxima and minima and if so find them.

(a) $y = x^2$;

(b) $y = 1 - 3x - x^2$;

(c) $y = x^3$;

(d) $y = x^3 - x$;

(e) $y = x^3 - 8$;

(f) $y = x^4 - x^2$;

(g) $y = x^2 + \dfrac{250}{x}$;   (h) $y = \sin x$;

(i) $y = 3 \sin x + 4 \cos x$;   (j) $y = x^2 \ln x$.

2. A cylindrical metal can with a lid is to have a specified volume V. Find the ratio of its height to its radius if the amount of metal used is to be a minimum.

3. The number of young which survive from a breeding colony of birds depends upon the density of nests within that colony. If the density is low, predators take a disproportionate toll and, if the density is high, food may be short and young can become lost or get killed by adults. A suggested relationship between the number N of young which survive and the density ρ of nests is

$$N = A\rho(B - \rho)$$

where A and B are constants. Find the optimum nest density if $A = 1000$ and $B = 4$.

4. A man is at sea in a rowing boat at a point P which is 3 km from the nearest point A of a straight shore line. He wishes to reach a point B, also on the shore, but at a distance of 5 km from A. The man can row at 2 kph and walk at 4 kph. At what point should he land (distance x from A) if he wishes to reach B in the least time? If his objective were a point Q, 2 km directly inland from B, what should be his path from P to Q to complete the journey in the least time?

5. Light travels in a straight line with speed v_1 in a medium m_1 and in a straight line with speed v_2 in a medium m_2. Show that in order for light to pass from a point P_1 in m_1 to a point P_2 in m_2 in the least time the angle of incidence θ_1 in m_1, measured from the normal to the interface between m_1 and m_2, and the angle of refraction θ_2 in m_2 are related to v_1 and v_2 by

$$\frac{\sin \theta_1}{\sin \theta_2} = \frac{v_1}{v_2}.$$

6. A population of size P is growing with time t according to

$$P = \frac{105}{1 + 20e^{-0.15t}}$$

where P is in millions and t is in days. Find an expression in terms of time for the rate R ($= dP/dt$) at which the population is growing. Hence determine at what time this rate of growth is at a maximum and the size of the population when this occurs.

4.7. Series expansions

There are many situations in which it would be useful if a rather complicated function, perhaps involving exponential or trigonometric expressions, could be replaced by something simpler such as a polynomial, even if this meant restricting its range of validity. It has been shown in §§3.6 and 3.7 that polynomials can be found which fit data very well and calculations based upon the polynomial are almost as accurate as those based upon the original function. Some polynomials exactly fit the data whereas others fit it well but not exactly. Increasing the degree of the polynomial often increases the accuracy. A quadratic interpolation is normally better than a linear interpolation (§3.7). It might therefore be expected that a polynomial of high degree can be found to fit data or represent a function to a greater accuracy than one of lower degree. This gives rise to the idea of *power series* which consist of a sequence of increasing powers of a variable, each with a coefficient, added together to form a series. If the variable is x such a series would be

$$a_0 + a_1x + a_2x^2 + a_3x^3 + \ldots$$

where the a_i, $i = 1, 2, 3 \ldots$, are coefficients and the series may be infinite in length. This possibility of infinite length distinguishes it from a polynomial which has a specified degree. It is known that some functions can be expanded in the form of a power series and a simple example is

$$\frac{1}{1-x} = 1 + x + x^2 + x^3 + \ldots$$

provided $|x| < 1$. The right-hand side is an infinite geometric progression (Appendix A.4) which can be summed to give the left-hand side provided $|x| < 1$. It is obvious that the two sides of the equation are equal at $x = 0$ and just as obvious that they are not equal if $x = 2$. In this latter case the left-hand side is -1 whereas the right-hand side consists of a series of positive terms which rapidly increase. Provided x is small the difference between the left-hand side and the first few terms of the right-hand side is very small. For example if $x = 0.1$ the difference between the left-hand side of $1\frac{1}{9}$ and the first three terms of the right-hand side giving 1.11 is less than 0.1 per cent. If increased accuracy were required the first four or five terms might be used. It would be useful to know whether such series representations of functions are always possible or whether the above example is rather unique.

In the most general situation we would like to know if $f(x)$ has a power series expansion and if so whether we can say just what it is. Let us assume that such an expansion is possible so that

$$f(x) = a_0 + a_1 x + a_2 x^2 + a_3 x^3 + \ldots.$$

This expansion is certainly true at $x = 0$ provided a_0 is chosen to equal $f(0)$. With this provision it seems likely that a series does exist for values of x near zero since each term then becomes less and less significant. Other coefficients can in fact be found using systematic differentiation.

$$f'(x) = a_1 + 2a_2 x + 3a_3 x^2 + \ldots.$$

Putting $x = 0$ again gives $a_1 = f'(0)$.

$$f''(x) = 2a_2 + 3 \times 2 \times a_3 x + \ldots$$

and putting $x = 0$ yet again shows that $a_2 = \frac{1}{2} f''(0)$. Similarly,

$$a_3 = \frac{1}{3 \cdot 2 \cdot 1} f'''(0),$$

$$a_4 = \frac{1}{4 \cdot 3 \cdot 2 \cdot 1} f''''(0),$$

$$\text{etc.}$$

In general $a_i = (1/i!) f^{(i)}(0)$ where $i! = 1 \times 2 \times 3 \times \ldots \times i$. ($i$ factorial) and $f^{(i)}(0)$ is the ith derivative of $f(x)$ evaluated at $x = 0$. The following general series expansion results

$$f(x) = f(0) + x f'(0) + \frac{x^2}{2!} f''(0) + \frac{x^3}{3!} f'''(0) + \ldots$$

and is known as the *Maclaurin series*. It is certainly valid at $x = 0$ and may be valid for a range of x including $x = 0$ or for all values of x depending upon the form $f(x)$ takes. Ways of determining the range of validity or range of *convergence* are beyond the scope of this book.

Example. Find the Maclaurin series expansion of $(1 + x)^3$.

The brackets can be multiplied out to give

$$(1 + x)^3 = 1 + 3x + 3x^2 + x^3.$$

The same result should come from the Maclaurin expansion.

$$f(x) = (1 + x)^3 \qquad \therefore \qquad f(0) = 1$$
$$f'(x) = 3(1 + x)^2 \qquad\qquad f'(0) = 3$$
$$f''(x) = 6(1 + x) \qquad\qquad f''(0) = 6$$
$$f'''(x) = 6 \qquad\qquad f'''(0) = 6$$
$$f^{(4)}(x) = 0 \qquad\qquad f^{(4)}(0) = 0$$
$$\text{etc.} \qquad\qquad\qquad \text{etc.}$$

All subsequent derivatives are zero.

Therefore, $\quad f(x) = (1 + x)^3 = 1 + 3x + \dfrac{x^2}{2!} \times 6 + \dfrac{x^3}{3!} \times 6$

$$= 1 + 3x + 3x^2 + x^3,$$

which checks with direct multiplication.

Example. Find the Maclaurin expansion of $\sin x$.

$$f(x) = \sin x \qquad\qquad f(0) = 0$$
$$f'(x) = \cos x \qquad\qquad f'(0) = 1$$
$$f''(x) = -\sin x \qquad\qquad f''(0) = 0$$
$$f'''(x) = -\cos x \qquad\qquad f'''(0) = -1$$
$$f^{(4)}(x) = \sin x \qquad\qquad f^{(4)}(0) = 0$$

and the sequence repeats itself. Therefore,

$$\sin x = x - \frac{x^3}{3!} + \frac{x^5}{5!} - \frac{x^7}{7!} + \ldots$$

Similarly,

$$\cos x = 1 - \frac{x^2}{2!} + \frac{x^4}{4!} - \frac{x^6}{6!} + \ldots$$

and

$$e^x = 1 + x + \frac{x^2}{2!} + \frac{x^3}{3!} + \frac{x^3}{4!} + \frac{x^4}{5!} + \ldots .$$

It can be shown that the series for $\sin x$, $\cos x$, and e^x are valid or convergent for all values of x. These series show that if x is small the approximations $\sin x \approx x$ and $\cos x \approx 1$ apply. These are in fact both *linear approximations* coming from the first two general terms of the Maclaurin series,

$$f(x) \approx f(0) + xf'(0),$$

provided x is small. A *quadratic approximation* would be

$$f(x) \approx f(0) + xf'(0) + \tfrac{1}{2}x^2 f''(0).$$

Example. Find an approximation to $\sqrt{10}$ using both the linear and quadratic approximations of the Maclaurin expansion.

We know $\sqrt{9} = 3$ and so let $f(x) = \sqrt{(9 + x)}$ in order that $f(1) = \sqrt{10}$.

$$f(x) = \sqrt{(9 + x)} \quad , \qquad f(0) = 3$$
$$f'(x) = \tfrac{1}{2}(9 + x)^{-\frac{1}{2}} \quad , \qquad f'(0) = \tfrac{1}{6}$$
$$f''(x) = -\tfrac{1}{4}(9 + x)^{-\frac{3}{2}} , \qquad f''(0) = -\tfrac{1}{108} .$$

A linear approximation is $\sqrt{(9 + x)} \approx 3 + (x/6)$ giving $\sqrt{10} \approx 3.1667$.

A quadratic approximation is $\sqrt{(9 + x)} \approx 3 + (x/6) - (x^2/216)$ giving $\sqrt{10} \approx 3.1621$.

The value from tables is 3.1623. Both results could have been obtained from the binomial expansion of $\sqrt{(9 + x)}$ written as $3(1 + \tfrac{1}{9}x)^{\frac{1}{2}}$ using the form for a fractional or negative index (Appendix A.2).

Example. A decaying population which exhibits annual oscillations can be represented by

$$P = P_0 \{3 + \sin(2\pi t)\} e^{-\frac{t}{10}}$$

where P_0 is a constant and t is time measured in years. Find a quadratic approximation to P for values of t near $2\tfrac{1}{2}$.

We may write $t = 2.5 + x$ so that

$$P = P_0 \{3 + \sin(2\pi(2.5 + x))\}e^{-\frac{2.5+x}{10}}$$

$$= P_0 e^{-0.25} \{3 - \sin(2\pi x)\}e^{-\frac{x}{10}}$$

after using the formula for $\sin(\phi + \theta)$ from Appendix B.1. The Maclaurin expansion of P in terms of x can now be found. Define $f(x)$ by

$$f(x) = \{3 - \sin(2\pi x)\}e^{-\frac{x}{10}}$$

so that

$$P = P_0 e^{-0.25} f(x).$$

$$f(x) = \{3 - \sin(2\pi x)\}e^{-\frac{x}{10}}, \qquad f(0) = 3.$$

$$f'(x) = \left\{ -\frac{3}{10} + \frac{1}{10}\sin(2\pi x) - 2\pi \cos(2\pi x) \right\}e^{-\frac{x}{10}},$$

$$f'(0) = -\left(2\pi + \frac{3}{10}\right).$$

$$f''(x) = \left\{ \frac{3}{100} + \left(4\pi^2 - \frac{1}{100}\right)\sin(2\pi x) + \frac{2\pi}{5}\cos(2\pi x) \right\}e^{-\frac{x}{10}},$$

$$f''(0) = \frac{1}{5}\left(2\pi + \frac{3}{20}\right).$$

Hence

$$f(x) \approx 3 - \left(2\pi + \frac{3}{10}\right)x + \frac{1}{10}\left(2\pi + \frac{3}{20}\right)x^2 + \dots$$

and so

$$P \approx P_0 e^{-0.25} \left\{3 - \left(2\pi + \frac{3}{10}\right)x + \frac{1}{10}\left(2\pi + \frac{3}{20}\right)x^2\right\}.$$

This series expansion can also be obtained by expanding the component parts of $f(x)$.

$$f(x) = \{3 - \sin(2\pi x)\}e^{-\frac{x}{10}}$$

$$= \left\{3 - \left(2\pi x - \frac{(2\pi x)^3}{3!} + \ldots\right)\right\}\left(1 - \frac{x}{10} + \frac{(x/10)^2}{2!} + \ldots\right)$$

$$\approx \{3 - 2\pi x\}\left(1 - \frac{x}{10} + \frac{x^2}{200}\right)$$

$$\approx 3 - \left(2\pi + \frac{3}{10}\right)x + \frac{1}{10}\left(2\pi + \frac{3}{20}\right)x^2.$$

If x is replaced by $t - 2.5$ we have

$$P \approx P_0 e^{-0.25}\left\{3 - \left(2\pi + \frac{3}{10}\right)\left(t - \frac{5}{2}\right) + \frac{1}{10}\left(2\pi + \frac{3}{20}\right)\left(t - \frac{5}{2}\right)^2\right\}$$

which is the appropriate series for t near $2\frac{1}{2}$.

There is a more general form of series expansion known as *Taylor series*. In this case the expansion can be with reference to any point, not just $x = 0$, and is a power series in $(x - a)$ where $x = a$ is the reference point. The series expansion is

$$f(x) = f(a) + (x - a)f'(a) + \frac{(x - a)^2}{2!}f''(a) + \frac{(x - a)^3}{3!}f'''(a) + \ldots$$

and reduces to the Maclaurin series if $a = 0$. An alternative form is

$$f(a + x) = f(a) + xf'(a) + \frac{x^2}{2!}f''(a) + \frac{x^3}{3!}f'''(a) + \ldots$$

obtained by replacing x by $a + x$ and hence $x - a$ by x.

The series expansions of $\sin x$, $\cos x$, and e^x give rise to a result which will be required later. The series expansion of e^x is

$$e^x = 1 + x + \frac{x^2}{2!} + \frac{x^3}{3!} + \frac{x^4}{4!} + \frac{x^5}{5!} + \frac{x^6}{6!} + \ldots$$

If into this is substituted $x = i\theta$ where i is a number such that $i^2 = -1$ (Appendix A.3) then we have

$$e^{i\theta} = 1 + i\theta - \frac{\theta^2}{2!} - \frac{i\theta^3}{3!} + \frac{\theta^4}{4!} + \frac{i\theta^5}{5!} + \dots$$

$$= \left(1 - \frac{\theta^2}{2!} + \frac{\theta^4}{4!} \dots\right) + i\left(\theta - \frac{\theta^3}{3!} + \frac{\theta^5}{5!} \dots\right)$$

$$= \cos\theta + i\sin\theta.$$

Hence the exponential and trigonometric functions are related. It can be shown in a similar manner that

$$e^{-i\theta} = \cos\theta - i\sin\theta$$

so that

$$\cos\theta = \frac{1}{2}(e^{i\theta} + e^{-i\theta}) \text{ and } \sin\theta = \frac{1}{2i}(e^{i\theta} - e^{-i\theta}).$$

More generally

$$e^{x+iy} = e^x(\cos y + i\sin y)$$

and so a *complex* exponential can be expressed partially in terms of trigonometric functions.

Exercises

1. Find the first three non-zero terms in the Maclaurin expansions of

(a) $\sin 2x$; (b) $\sqrt[3]{(1+x)}$; (c) $\cos^2 x$;

(d) $\tan x$; (e) $\ln(1+x)$.

2. Find the first five items in the Maclaurin expansion of

$$\left(1 + \frac{x}{n}\right)^n$$

where n is a positive integer. By observing what happens to the expansion as n becomes larger deduce that

$$\lim_{n \to \infty} \left(1 + \frac{x}{n}\right)^n = e^x.$$

3. By considering the first three terms of the Maclaurin expansion of $\log(10 + x)$ find approximations to log 9, log 9.7, log 10.5, and log 12. Compare these results with tabulated values. The following relationship may be required:

$$\log N = \frac{\ln N}{\ln 10} = \frac{\ln N}{2.30259}.$$

4. A population of size P depends upon time t according to

$$P = \frac{50}{1 + 4e^{-0.1t}}.$$

Find a quadratic approximation to this expression valid near $t = 0$. Use trial and error to find approximately how large t can be before the quadratic differs from the exact equation by more than 10 per cent.

4.8. Small changes and errors

The derivative $f'(x)$ is defined by (§4.3)

$$f'(x) = \lim_{\delta x \to 0} \frac{f(x + \delta x) - f(x)}{\delta x}$$

and so, provided δx is sufficiently small,

$$f'(x) \approx \frac{f(x + \delta x) - f(x)}{\delta x}.$$

Hence

$$f(x + \delta x) \approx f(x) + f'(x)\delta x.$$

Note that this is the same as a linear approximation to $f(a + x)$ using Taylor series if a is replaced by x and x by δx. This approximation can be used to estimate the new value of $f(x)$ when x is increased by δx. The resultant change, δf, in the function is given by

$$\delta f = f(x + \delta x) - f(x) \approx f'(x)\delta x$$

or

$$\delta f \approx \frac{\mathrm{d}f}{\mathrm{d}x}\delta x.$$

The smaller the change δx the greater will be the accuracy of this approximation.

Example. A population P is believed to be growing according to the equation

$$P = 10\,000\,e^{0.043t}$$

where t is the time in hours. At a certain time the population is 15 000. What is the approximate increase over the next half hour?

Let the time at which the population is 15 000 be T so that

$$15\ 000 = 10\ 000\ e^{0.043T}.$$

Now

$$\delta P \approx \frac{dP}{dt}\delta t = 430\ e^{0.043t}\delta t.$$

At time T, $e^{0.043T} = 1.5$ and to cover the next half hour $\delta t = 0.5$. The change in the population is therefore approximately

$$\delta P \approx 430 \times 1.5 \times 0.5 \approx 322.$$

(The exact answer is 326.00).

The same process can be used to find errors in the dependent variable resulting from errors in the independent variable.

Example. The rate of flow of blood through an artery is proportional to the fourth power of the internal diameter of the artery. If measurements of this diameter are subject to an error of ±7 per cent, find the resultant error in the rate of flow.

Since the rate of flow is proportional to d^4 where d is the diameter let $R = kd^4$ where R is the rate and k a constant of proportionality. We are given that

$$\delta d/d = \pm\frac{7}{100}.$$

$$\delta R \approx \frac{dR}{dd}\delta d = 4kd^3\delta d.$$

Therefore, $$\frac{\delta R}{R} \approx \frac{4kd^3\delta d}{kd^4} = 4\frac{\delta d}{d} = \pm\frac{28}{100}.$$

The resultant error in the flow is therefore ±28 per cent.

Exercises

1. A metal cube of side 10 cm is electroplated with chromium. If the thickness of the chromium is 0.5 mm, find approximately the volume deposited.

2. The periodic time T (s) of a simple pendulum of length l (m) in a gravitational field g (m s^{-2}) is given by

$$T = 2\pi\sqrt{(l/g)}.$$

If there is a 2.5 per cent error in l what is the resultant error in T?

3. Estimate $\sqrt[3]{1002}$ using $y = \sqrt[3]{x}$ with $x = 1000$ and $\delta x = 2$. Use the same method to find $\sqrt[3]{990}$.

4. For a fixed concentration of a given solution the intensity I of light at a depth of penetration x (cm) is given by

$$I = I_0 e^{-0.137x}$$

where I_0 is the initial intensity. If the intensity has fallen to 50.4 per cent of its original value at the depth of 5 cm use differentiation to find approximately how much further it falls at a depth of 5.5 cm.

5. A population oscillates with a 10-year period according to

$$P = 1.5 - 0.5 \cos(0.2\pi t)$$

where P is in millions and t in years. Find an expression for the rate of change of the population as a function of P alone. Hence find the approximate increase in the population over the next three months if at a certain time it is 1.4 million and rising.

4.9. Related rates

The chain rule gives a method of finding the rate of change of the dependent variable with respect to a new variable if the rate of change of the independent with respect to this new variable is known. It will be assumed that the new variable is time t but this will not necessarily always be the case. If

$$y = f(x),$$

then

$$\frac{dy}{dt} = \frac{dy}{dx} \cdot \frac{dx}{dt} = f'(x)\frac{dx}{dt}.$$

Example. A spherical balloon is being filled with helium at a constant rate of 5 litres per second. If it is assumed that the balloon remains spherical during the filling process and the pressure inside remains constant find the rate at which its radius is increasing after 2000 litres have passed in.

The volume of the sphere after 2000 litres have passed in is 2 m^3. The rate of filling at 5 litres per second is 0.005 m^3 s^{-1}. The volume V of a sphere of radius r is given by

$$V = \frac{4}{3}\pi r^3.$$

Hence,
$$r = \left(\frac{3V}{4\pi}\right)^{\frac{1}{3}}.$$

Therefore,
$$\frac{dr}{dt} = \frac{dr}{dV} \cdot \frac{dV}{dt}$$

$$= \left(\frac{3}{4\pi}\right)^{\frac{1}{3}} \frac{1}{3} V^{-\frac{2}{3}} \frac{dV}{dt} = \frac{1}{\sqrt[3]{(36\pi V^2)}} \frac{dV}{dt}$$

$$= \frac{1}{\sqrt[3]{(36\pi \times 4)}} \times 0.005 = 6.5 \times 10^{-4}\,\text{m s}^{-1}.$$

The rate of increase of the radius at this stage is therefore 0.065 cm s^{-1}. The radius itself is 78.16 cm.

Example. The gross photosynthetic rate R of a plant is related to the rate at which light energy E falls on it by

$$R = \frac{1}{4.0 + (0.1/E)}.$$

Find the rate at which R is changing when E is 0.08 W m^{-2} and falling at a rate of 0.02 W m^{-2} per hour.

Now
$$R = \frac{E}{4.0E + 0.1}$$

and differentiating with respect to time gives

$$\frac{\mathrm{d}R}{\mathrm{d}t} = \frac{\mathrm{d}R}{\mathrm{d}E} \cdot \frac{\mathrm{d}E}{\mathrm{d}t}$$

$$= \frac{(4E + 0.1) - E \times 4}{(4E + 0.1)^2} \frac{\mathrm{d}E}{\mathrm{d}t}$$

$$= \frac{0.1}{(4E + 0.1)^2} \frac{\mathrm{d}E}{\mathrm{d}t}.$$

At the time in question $E = 0.08$ and $\mathrm{d}E/\mathrm{d}t = -0.02$, the minus sign being included because the light energy rate is decreasing.

Therefore, $$\frac{\mathrm{d}R}{\mathrm{d}t} = \frac{0.1}{(4 \times 0.08 + 0.1)^2} \times (-0.02)$$

$$= -0.011 \text{ units per hour.}$$

The photosynthetic rate is therefore decreasing at the rate of 0.011 units per hour.

Exercises

1. A gas is initially at atmospheric pressure and at this pressure occupies a volume of 1 m³. The pressure however is being steadily increased at the rate of two atmospheres per hour. If it is assumed that the product of pressure and volume is constant, find the rate at which the volume is changing after two hours.

2. The focal length f of the human eye can be controlled so that an object at almost any distance u in front of the lens can be brought into focus on the retina at a distance $v = 2.5$ cm behind the lens. Assuming that the eye obeys the thin lens formula,

$$\frac{1}{u} + \frac{1}{v} = \frac{1}{f},$$

find the rate at which the focal length of the lens must be changing to keep in focus an object moving towards the eye at a constant speed of 0.7 m s^{-1} when it is at a distance of 3.0 m from the eye.

3. The mean rate R at which radiant energy falls on a pond in June is $650 \sin \theta$ W m^{-2} where θ is the angle of elevation of the sun above the horizon. If the sun's angle of elevation in June is given by $\frac{1}{3}\pi \sin(0.205t)$ radians, where t is the time in hours after sunrise, find the rate at which R is changing when $t = 1, 3, 8,$ and 12 hours.

4.10. The Newton-Raphson method

Before concluding this chapter on differentiation mention should be made of a popular and very effective method of solving equations which depends upon a derivative. The solution of polynomial equations of degrees 1, 2, 3, and 4 can be achieved with increasing difficulty. Beyond this only certain forms can be solved by analytic methods. For equations which are not polynomial in form it is a matter of luck whether the particular equation encountered can be solved exactly at all. Equations which are a combination of algebraic and trigonometric functions for instance almost certainly cannot be solved by analytic means.

For the more difficult equations approximate methods of solution have to be used. These normally require an initial estimate of the solution which is then fed into some mathematical process to give a second but better estimate. The second estimate is then fed in to produce a third estimate and the process is repeated as many times as required to obtain a specified degree of accuracy. Because such methods are repetitive and often quite simple they are very well suited to being programmed on a computer. They are known as *iterative* methods, each estimate being called an *iterate*. They are a branch of *numerical methods* used for problems when exact mathematical or *analytical* methods cannot be applied or are not known.

Each iterative method requires an intial estimate to start with and a set process or *algorithm* which is then applied to it to improve upon it. The initial estimate can be obtained in numerous ways. If the roots (solutions) of $f(x) = 0$ are required the points at which a graph of $y = f(x)$ cross the x-axis can provide a fairly accurate starting point. A piece of inspired guesswork can be equally as good. Some methods require quite a good initial estimate whereas others can work with what appears to be an impossibly bad one. The method to be described here comes into the second category which is why it is so popular. There are however circumstances in which it might fail or find an alternative root to the one being sought. This is true of all methods.

It will be assumed that we wish to find the root (solution) of

$$f(x) = 0$$

given that an initial approximation is x_0 so that $f(x_0) \approx 0$. Denote the difference between the exact root x and the initial approximation x_0 by ϵ_0, i.e.

$$x - x_0 = \epsilon_0.$$

The Taylor expansion of $f(x)$ about x_0 (§4.7) then gives

$$f(x) = f(x_0) + (x - x_0)f'(x_0) + \frac{(x - x_0)^2}{2!}f''(x_0) + \frac{(x - x_0)^3}{3!}f'''(x_0) + \ldots$$

$$= f(x_0) + \epsilon_0 f'(x_0) + \frac{1}{2}\epsilon_0^2 f''(x_0) + \frac{1}{6}\epsilon_0^3 f'''(x_0) + \ldots$$

$$= 0$$

since the equation being solved in $f(x) = 0$. If the error ϵ_0 is not too large, a few terms of the Taylor series will suffice, the remaining terms being neglected as increasing powers of ϵ_0 make them insignificant. If this is assumed to be the case for terms involving ϵ_0^4 and higher powers then we have the equation

$$\frac{1}{6}\epsilon_0^3 f'''(x_0) + \frac{1}{2}\epsilon_0^2 f''(x_0) + \epsilon_0 f'(x_0) + f(x_0) = 0,$$

for the unknown error ϵ_0. In the simplest case we might neglect terms involving ϵ_0^2 and higher powers to give

$$\epsilon_0 f'(x_0) + f(x_0) = 0$$

and so

$$\epsilon_0 \approx -\frac{f(x_0)}{f'(x_0)}.$$

The approximately equals sign is used because terms in the Taylor series have been neglected and so we cannot expect this answer to be exact. It is however an approximation to the error which means that adding the approximation to the initial estimate x_0 will give an improvement. If this improved approximation is denoted by x_1 then

$$x_1 = x_0 - \frac{f(x_0)}{f'(x_0)}.$$

The process can now be repeated by letting the new error $x - x_1 = \epsilon_1$ and finding

$$\epsilon_1 \approx -\frac{f(x_1)}{f'(x_1)}.$$

An even better approximation to the root is therefore

$$x_2 = x_1 - \frac{f(x_1)}{f'(x_1)}.$$

In general we have the iterative formula or algorithm

$$x_{n+1} = x_n - \frac{f(x_n)}{f'(x_n)}$$

which enables an improved approximation x_{n+1} to be calculated from the current approximation x_n. It is known as the *Newton–Raphson method*.

What is happening geometrically is shown in Figure 4.13. An approximation x_n is known (initially $n = 0$) which lies reasonably close to the point at which the graph of $y = f(x)$ crosses the x-axis, the

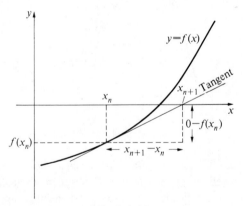

FIG. 4.13

exact root. The tangent to the graph at $x = x_n$ crosses the x-axis nearer to this root and provides a better approximation x_{n+1}. The slope of the tangent is $f'(x_n)$ and so from the figure,

$$f'(x_n) = \frac{0 - f(x_n)}{x_{n+1} - x_n}.$$

Therefore,
$$x_{n+1} = x_n - \frac{f(x_n)}{f'(x_n)}.$$

It can be shown that in general the rate at which the approximation improves at each iteration, its rate of *convergence*, is such that the number of correct significant figures doubles each time. The iterative formula does show however that there might be problems if $f'(x_n) \approx 0$. In this situation the graph of $y = f(x)$ is running almost parallel with the x-axis near the root. This can occur if there are two roots very close together (Figure 4.14). Between the two an approximation x_n might be found for which $f'(x_n) = 0$ and the method will fail. If $f'(x_n) \approx 0$, the rate of convergence is reduced compared with a more typical situation.

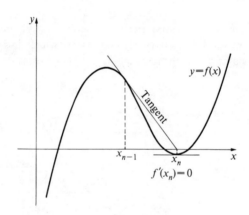

FIG. 4.14

Example. Derive an algorithm for finding the square root of a number, N, based upon the Newton-Raphson method and use it to find $\sqrt{10}$ to ten significant figures with 3 as an initial approximation.

We require to solve $x^2 = N$ for x. The equation must be expressed in the form $f(x) = 0$ and so

$$f(x) = x^2 - N = 0,$$
$$f'(x) = 2x.$$

Therefore,
$$x_{n+1} = x_n - \frac{f(x_n)}{f'(x_n)} = x_n - \frac{x_n^2 - N}{2x_n}$$
$$= \frac{1}{2}\left(x_n + \frac{N}{x_n}\right).$$

Using $N = 10$, $x_0 = 3$ and repeatedly applying the algorithm gives

$$x_0 = 3.000000000,$$
$$x_1 = 3.166666667,$$
$$x_2 = 3.162280702,$$
$$x_3 = 3.162277660,$$
$$x_4 = 3.162277660.$$

The last two iterates are identical and we can conclude that $x_3 = 3.162277660$ is $\sqrt{10}$ to 10 significant figures.

These methods are particularly well suited to a computer and the following is a program in BASIC to solve the equation

$$\sin 2x - x = 0$$

in which the program requests an initial approximation in radians. The expressions FNA(X) and FNB(X) are the function and its first derivative respectively. The program prints out each approximation, although this is not essential, and terminates when two successive approximations differ by less than 10^{-12}.

```
10  PRINT "NEWTON–RAPHSON FOR SIN(2 * X) – X = 0": PRINT
20  DEF FNA(X) = SIN(2 * X) – X: REM DEFINITION OF FUNCTION
30  DEF FNB(X) = 2 * COS(2 * X) – 1: REM DEFINITION OF DERIVATIVE
40  INPUT "INPUT INITIAL APPROXIMATION X0 = ", X: PRINT
50  PRINT "STEP N", "APPROX ROOT X"
60  PRINT 0, X
70  FOR N = 1 TO 50
80  X1 = X – FNA(X)/FNB(X)
90  PRINT N, X1
100 IF ABS(X1 – X) < 1E – 12 THEN 140
110 X = X1
120 NEXT N
130 PRINT: PRINT "SLOW CONVERGENCE": PRINT: GOTO 150
140 PRINT: PRINT "ROOT = "; X1, "STEPS TAKEN = "; N: PRINT
150 END
```

On running the program with an initial approximation of $x_0 = 1$ the following print out is obtained

> NEWTON–RAPHSON FOR SIN(2 * X) − X = 0
>
> INPUT INITIAL APPROXIMATION X0 = ?1
>
> | STEP N | APPROX ROOT X |
> |--------|---------------|
> | 0 | 1 |
> | 1 | .950497797103 |
> | 2 | .9477558226905 |
> | 3 | .947747133605 |
> | 4 | .9477471335176 |
> | 5 | .9477471335176 |
>
> ROOT = .9477471335176   STEPS TAKEN = 5
>
> END PROGRAM

A knowledge of the Newton–Raphson method should be enough to follow the program reasonably well. For the solution of a different equation statements 10, 20, and 30 should be appropriately changed.

Example. The size P of a population after t years is given by (cf. examples in §4.5 and §4.6)

$$P = 5000 \left(1 + \frac{1}{5}\sin(2\pi t) + \frac{1}{2}\sin\left(\frac{2\pi}{11}t\right)\right).$$

Find the time at which the first local maximum of this population occurs and the population size at this time.

Equating the derivative dP/dt to zero leads to the equation (cf. example in §4.6)

$$22\cos(2\pi t) + 5\cos\left(\frac{2\pi}{11}t\right) = 0.$$

The form of the expression for P indicates that its first local maximum occurs near the first maximum of $\sin(2\pi t)$. This occurs when $t = \frac{1}{4}$. We therefore define a function $f(t)$ by

$$f(t) = 22\cos(2\pi t) + 5\cos\left(\frac{2\pi}{11}t\right)$$

and use the Newton–Raphson method to solve $f(t) = 0$. For this the derivative,

$$f'(t) = -2\pi\left(22\ \sin(2\pi t) + \frac{5}{11}\sin\left(\frac{2\pi}{11}t\right)\right),$$

is required.

The program above can be used provided the appropriate changes are made. The first three statements should read

```
10 PRINT "NEWTON-RAPHSON FOR 22 * COS(2 * #PI * X) +
   5 * COS(2 * #PI * X/11) = 0": PRINT
20 DEF FNA(X) = 22 * COS(2 * #PI * X) + 5 * COS(2 * #PI * X/11): REM
   DEFINITION OF FUNCTION
30 DEF FNB(X) = −2 * #PI * (22 * SIN(2 * # PI * X) + 5 * SIN(2 * #PI *
   X/11)/11): REM DEFINITION OF DERIVATIVE
```

Using an initial approximation of X0 = 0.25 the following print out is obtained when the program is run.

NEWTON-RAPHSON FOR 22 * COS(2 * #PI * X) + 5 * COS(2 * #PI * X/11) = 0

INPUT INITIAL APPROXIMATION X0 = ? .25

STEP N	APPROX ROOT X
0	.25
1	.2856984361698
2	.2859960382363
3	.2859961011686
4	.2859961011687

ROOT = .2859961011687   STEPS TAKEN = 4

This root is the time in years at which the first critical point of P occurs. The derivatives of P are

$$\frac{dP}{dt} = 10000\pi\left(\frac{1}{5}\cos(2\pi t) + \frac{1}{22}\cos\left(\frac{2\pi}{11}t\right)\right)$$

and

$$\frac{d^2P}{dt^2} = -20000\pi^2\left(\frac{1}{5}\sin(2\pi t) + \frac{1}{242}\sin\left(\frac{2\pi}{11}t\right)\right).$$

At the root given the first derivative vanishes and the second is negative and so there is a maximum. The value of this first maximum is $P = 6381$ and it occurs after 0.286 years (104 days). If the starting point is the beginning of a calendar year then the maximum occurs in mid April of the first year.

Exercises

1. Using graphs or otherwise find approximate solutions to the following equations and improve upon them by two applications of the Newton–Raphson method.

(a) $x^2 = \sin x$;   (b) $xe^x = \cos x$;   (c) $x^2 + \ln x = 0$.

2. Two populations, P_1 and P_2, each in millions, are given by,

$$P_1 = 4 - 2\cos(0.1t), \quad P_2 = \frac{6}{1 + 5e^{-0.2t}},$$

where t is time in years. At what time are these two populations equal?

3. The expected yield Y per potato plant in kg can be expressed as a function of the planting density ρ, where ρ is the number of plants per hectare. Two such yield expressions are

$$Y_1 = \frac{10^6}{8.5\rho + 40000} \quad \text{and} \quad Y_2 = 25\left(1 - e^{-\frac{3850}{\rho}}\right).$$

Determine where these two yield functions are equal.

5 Integration

5.1. The anti-derivative

IN THE PREVIOUS chapter the derivative has been defined, methods of differentiation explained, and wide areas of application demonstrated. With such importance attaching to rates of change in numerous areas of the biological sciences it might be wondered whether the complete form of a function can be deduced from a knowledge of its derivative. Is it possible to perform an anti-differentiation to produce an *anti-derivative*? There are certainly areas where this would be useful, areas where just a rate of change is known.

Simple population growth has been described in §2.6 where it was shown to be exponential in form.

$$P = P_0 2^{2t} = P_0 e^{2t \ln 2}.$$

More generally

$$P = P_0 e^{kt}$$

where k is a positive constant. The rate of change of such a population is given by

$$\frac{\mathrm{d}P}{\mathrm{d}t} = k P_0 e^{kt} = kP.$$

In other words the rate of change of the population is proportional to its size at any time. Such growth increases without limit which is not possible in a real-life situation. There will be some upper bound beyond which the population will not grow. As it approaches this bound the population will tend to level out, that is its rate of increase will be less steep than previously. In saying this we are describing the behaviour of the derivative. It must remain positive but get smaller and smaller as the upper bound on the population is approached. The simplest function to do this is the linear function $(P_M - P)$ where P_M represents the upper bound on the population and P is the actual population. The rate of change of the population may therefore be described by

$$\frac{\mathrm{d}P}{\mathrm{d}t} = k(P_M - P)$$

where k is a positive constant of proportionality. Note that if P ever reaches P_M then $dP/dt = 0$ and growth stops. Note also that if the population ever exceeds P_M, due to a migratory influx perhaps, then dP/dt is negative and the population decreases towards P_M. In all cases therefore the equation above describes quite well the type of behaviour that might be expected from a population of sufficient size to be affected by its limited environment. The equation describes the rate of change of the population rather than the population itself. It contains a derivative and is an example of a *differential equation*. It is in situations like this that an anti-derivative would be useful. We could then find the population itself as a function of time.

In seeking the anti-derivative of the function $f(x)$, say, we are seeking a function $F(x)$ whose derivative with respect to x is $f(x)$. Given

$$\frac{dy}{dx} = f(x),$$

we require

$$y = F(x).$$

In terms of time, distance, speed, and acceleration it is as if we knew the acceleration as a function of time but required the speed, or knew the speed at any time but wished to find the distance travelled. Let us assume that we know the acceleration a as a function of time,

$$a = a(t).$$

Now acceleration is the time derivative of speed and so

$$\frac{dv}{dt} = a(t).$$

Let us also assume that by careful monitoring of the situation we can work out, using the fact that acceleration is the rate of change of speed, what the total change in speed has been from time 0 to some arbitrary time t. Denoting this change by $v(t)$, since it will certainly depend upon the value of t chosen, we might think that the speed at any time t is given by

$$v = v(t).$$

This is not necessarily the case because by monitoring just the acceleration over the period we have only found the change in speed, not the absolute speed. If $t = 0$, we have

$$v = v(0) = 0,$$

since the time period over which the change was monitored is now zero and hence the change in speed is also zero. In order that v should represent the absolute speed we must also know what the speed was at the beginning of the period. If this speed, the initial speed, is v_0, then

$$v = v_0 + v(t)$$

where v now represents the absolute speed. This value of v_0 could not be determined by monitoring just acceleration. It has to be known independently.

The careful monitoring of acceleration has amounted to finding its anti-derivative, the speed. Differentiation should therefore give back the acceleration.

$$\frac{dv}{dt} = \frac{d}{dt}(v_0 + v(t)) = \frac{dv_0}{dt} + \frac{dv(t)}{dt}$$

$$= 0 + a(t).$$

The derivative of the initial speed, a constant, is zero and so the acceleration is recovered. The point to note is that whatever this initial speed was it does not appear at all after differentiation to obtain the acceleration. Conversely, anti-differentiation, a process applied solely to the expression for acceleration, can tell us nothing about the initial speed.

The same argument can be applied to a situation in which the speed $v = v(t)$ is known as a function of time. Anti-differentiation will give the change in distance $x(t)$ between some zero time and an arbitrary time t. The absolute distance x would, however, require the addition of the distance x_0 already travelled from some origin prior to the zero time. Hence,

$$\frac{dx}{dt} = v(t)$$

and

$$x = x_0 + x(t).$$

Again there is no information in the speed function $v(t)$ from which this constant x_0 could have been calculated.

In general terms, therefore, if

$$\frac{dy}{dx} = f(x)$$

and the anti-derivative $F(x)$ can be found, an arbitrary constant should be added to it to give

$$y = F(x) + C.$$

This constant cannot be determined from the derivative dy/dx. Some further independent information is required to find a value for it. Differentiation is a destructive process, constants vanish without trace, and to allow for this the arbitrary constant must be added to the anti-derivative.

As far as the anti-derivative itself is concerned we know more than we might think. In the previous chapter quite a wide range of derivatives were calculated and looking at each of these in reverse gives an anti-derivative. We know for example that the derivative of x^n is nx^{n-1}. Conversely therefore the anti-derivative of nx^{n-1} is x^n. Tidying this up by replacing n by $n + 1$ and then dividing by $n + 1$ shows that the anti-derivative of x^n is $x^{n+1}/(n + 1)$. The result can be checked by differentiation which gives back x^n. Using this process a short table can be produced.

Function $f(x)$	Anti-derivative $F(x)$
x^α	$\dfrac{x^{a+1}}{\alpha + 1}, \quad \alpha \neq -1$
$\dfrac{1}{x}$	$\ln x$
e^x	e^x
a^x	$\dfrac{a^x}{\ln a}, \quad a > 0, a \neq 1.$
$\sin x$	$-\cos x$
$\cos x$	$\sin x$

In the first result $\alpha = -1$ is excluded since the value would mean
dividing through by zero. The $\alpha = -1$ situation is covered by the
second result. In the fourth result the restrictions on a are those that
apply to any base.

5.2. Area under a curve

One property associated with a point on a curve is the slope of the
curve represented by the slope of the tangent to the curve at that point.
The significance of this slope has been demonstrated in §4.1 where, for
a distance against time curve, it gave the instantaneous speed. Another,
although cumulative, property of a point on a curve is the area between
the curve and the axes. For a curve as shown in Figure 5.1, this
property is cumulative in the sense that it depends upon previous points

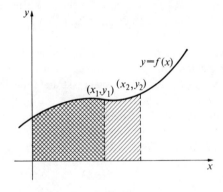

F IG . 5.1

through the slope of the curve to the left and increases as the point
itself moves to the right. This area increases as x increases.

Using time, distance, and speed to illustrate the situation consider a vehicle moving with a constant speed of 30 m s^{-1}. A graph of
speed v against time t is a horizontal straight line as shown in Figure
5.2. Since speed is constant the distance travelled, x, in time t is
given by

$$x = 30t.$$

The area between the line $v = 30$ and the horizontal axis ($v = 0$),
bounded on the left by the vertical axis ($t = 0$) and on the right by the

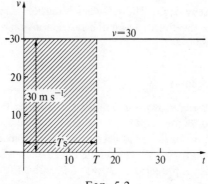

F IG . 5.2

time point $t = T$, has dimensions of speed multiplied by time. It therefore represents a distance and furthermore the area is numerically equal to $30T$ (Figure 5.2), the exact distance travelled from time $t = 0$ to time $t = T$.

It was shown in §4.2 that a free-falling stone had speed $v = 10t$ for an assumed gravitational acceleration of 10 m s^{-2}. The speed against time graph, shown in Figure 5.3, is again a straight line but in this case it slopes upwards. The area beneath it from $t = 0$ to $t = T$ can be calculated using the area for a triangle formula of half the base (time) multiplied by the vertical height (speed). Dimensionally this area represents a distance. If the speed is V when $t = T$ then $V = 10T$ and numerically the area is

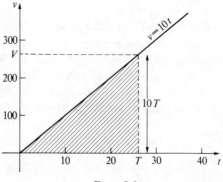

F IG . 5.3

$$\frac{1}{2} \times T \times V = \frac{1}{2} \times T \times 10T = 5T^2.$$

Recalling from §4.1 that the speed $v = 10t$ was derived from a distance equation $x = 5t^2$ we can see that the area calculated above has given exactly this distance when $t = T$.

In each of these two examples the area has been measured from the vertical axis, $t = 0$. If we required the distance travelled from $t = 5$ to $t = 20$, say, in each of these cases we would calculate the distance at $t = 20$ and subtract from it the distance at $t = 5$. Diagramatically this is calculating a larger area and then cutting off part of it, leaving the area below the curve between $t = 5$ and $t = 20$ as shown in Figure 5.4.

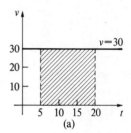 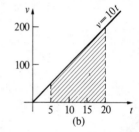

FIG. 5.4

In the case of

$$x = x(t) = 30t$$

the two distances are

$$x(5) = 150,$$
$$x(20) = 600,$$

giving a distance travelled of 450 metres. This is exactly the area below $v = 30$ from $t = 5$ to $t = 20$. For the second case where

$$x = x(t) = 5t^2,$$

we have

$$x(5) = 125,$$
$$x(20) = 2000,$$

giving a distance of 1875 metres. The speed against time graph of $v = 10t$ and the area involved is shown in Figure 5.4(b). It can be divided into a triangle upon a rectangle, the dimensions of which are shown in Figure 5.5. The triangle has area 1125 units and the rectangle 750 units giving 1875 units altogether, the units being m s^{-1} × s or metres.

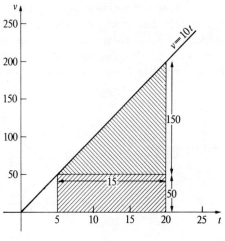

F IG . 5.5

In general the area under any curve will have dimensions given by the product of the units used along each of the coordinate axes. The area under a graph of blood-flow rate (litres per second) against time (seconds) will have dimensions of total blood flow (litres). The two examples above show that the magnitude of the area has a significance and there is a strong inclination to believe that generally this is the case also. However we have only considered examples in which the 'curve' involved was a straight line because areas are then easy to calculate. A more detailed and systematic approach is required if the area under a less simple graph is required.

The area under any curve can be found approximately by dividing it up into a large number of thin vertical strips. The top end of each strip will take on the shape of the curve in that region. The area of such a strip will not be significantly different from a similar strip in which the curved tip has been snipped off to make it level. This second strip is now a rectangle and its area is simple to calculate. The total area of all such rectangles will be very close to the area under the curve and the narrower the strips the less will be the error. Such a situation is shown in Figure 5.6.

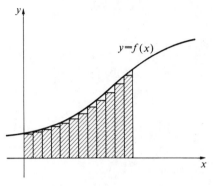

FIG. 5.6

To see how well the method works it will be applied to the area under $v = 10t$ since this area can be calculated as previously and therefore provides a check. The area to be found is shown in Fig. 5.7 and covers the period from $t = 5$ to $t = 20$. This 15-second interval is divided into n equal steps to provide n strips each of which is levelled off at the top. The height of the first strip is given by

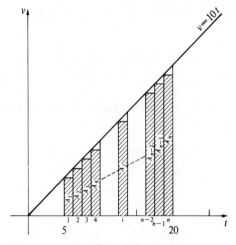

FIG. 5.7

$$v_1 = 10 \times 5$$

and since the strip is $15/n$ wide the area is

$$A_1 = \frac{15}{n} \times v_1 = \frac{15}{n} \times 10 \times 5.$$

The height of the second strip is

$$v_2 = 10 \times \left(5 + \frac{15}{n}\right)$$

and its area

$$A_2 = \frac{15}{n} \times v_2 = \frac{15}{n} \times 10 \times \left(5 + \frac{15}{n}\right).$$

The third strip has height

$$v_3 = 10 \times \left(5 + 2 \times \frac{15}{n}\right)$$

and area

$$A_3 = \frac{15}{n} \times v_3 = \frac{15}{n} \times 10 \times \left(5 + 2 \times \frac{15}{n}\right)$$

The ith strip has height

$$v_i = 10 \times \left(5 + (i-1) \times \frac{15}{n}\right)$$

and area

$$A_i = \frac{15}{n} \times v_i = \frac{15}{n} \times 10 \times \left(5 + \frac{15}{n}(i-1)\right).$$

An approximation to the whole area is therefore the sum of all n such areas. This is given by

$$A = \sum_{i=1}^{n} \frac{150}{n} \left(5 + \frac{15}{n}(i-1)\right)$$

$$= \frac{150}{n} \left(\sum_{i=1}^{n} 5 + \frac{15}{n} \sum_{i=1}^{n}(i-1)\right)$$

$$= \frac{750}{n} \sum_{i=1}^{n} 1 + \frac{2250}{n^2} \sum_{i=1}^{n} (i-1).$$

Now

$$\sum_{i=1}^{n} 1 = 1 + 1 + 1 + \ldots + 1, n \text{ times} = n$$

and

$$\sum_{i=1}^{n} (i-1) = 0 + 1 + 2 + 3 + \ldots + (n-2) + (n-1).$$

The sum of the first n integers is

$$1 + 2 + 3 + \ldots + (n-1) + n = \frac{1}{2} n(n+1)$$

(Appendix A.4) and so the second summation above is the sum of the first $(n-1)$ integers. Hence

$$\sum_{i=1}^{n} (i-1) = \frac{1}{2} (n-1)n$$

and so

$$A = \sum_{i=1}^{n} A_i = \frac{750}{n} \times n + \frac{2250}{n^2} \times \frac{1}{2} (n-1)n$$

$$= 750 + 1125 \times \left(1 - \frac{1}{n} \right).$$

The first term is the area of the rectangular portion and is exactly as before while the second is the approximation to the triangular portion. This clearly shows that the larger the number n of strips that the area is divided into the higher is the accuracy since the previous evaluation of this area was 1125 units. Since n has not been specified, we can apply the limit process and see what happens as n gets larger and larger.

$$\lim_{n \to \infty} \left(1 - \frac{1}{n}\right) = 1$$

and the exact result is obtained.

This gives some confidence in the method which can perhaps be applied to other areas. Let us consider the graph of

$$y = x^2$$

and attempt to calculate the area between $x = 2$ and $x = 5$ (Figure 5.8). A very rough approximation to the area can be found by drawing the chord from (2, 4) to (5, 25). The area below this chord is (3 × 4) + ($\frac{1}{2}$ × 3 × 21) = 43.5. The area below the curve is less than this so we might expect the answer to be about 40 in round figures. There are two

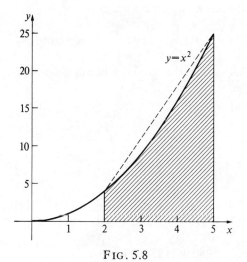

FIG. 5.8

ways of approaching the problem. The first is the direct approach used above in which the area between $x = 2$ and $x = 5$ is calculated immediately. The second is a more general approach in which the area from 0 to x, where x is not specified, is calculated. The area from 2 to 5 is then deduced from this by finding the area from $x = 0$ to $x = 5$ and subtracting the area from $x = 0$ to $x = 2$.

The second method will be used here and the strips are shown in Figure 5.9. Again n strips are chosen, this time each of width x/n.

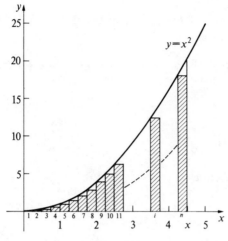

FIG. 5.9

The height of the ith strip is

$$y_i = \left\{ (i-1)\frac{x}{n} \right\}^2$$

and so

$$A_i = \frac{x}{n} \times y_i = \frac{x}{n} \left\{ (i-1)\frac{x}{n} \right\}^2 .$$

The total area is therefore

$$A = \sum_{i=1}^{n} A_i = \sum_{i=1}^{n} \frac{x}{n} \left\{ (i-1)\frac{x}{n} \right\}^2$$

$$= \sum_{i=1}^{n} \frac{x^3}{n^3} (i-1)^2$$

$$= \frac{x^3}{n^3} \sum_{i=1}^{n} (i-1)^2 .$$

Now

$$\sum_{i=1}^{n} (i-1)^2 = 0^2 + 1^2 + 2^2 + 3^2 + \ldots + (n-2)^2 + (n-1)^2$$

which is the sum of the squares of the first $(n - 1)$ integers. From Appendix A.4

$$1^2 + 2^2 + 3^2 + \ldots + (n - 1)^2 + n^2 = \frac{1}{6}n(n + 1)(2n + 1)$$

and so

$$\sum_{i=1}^{n} (i - 1)^2 = \frac{1}{6}(n - 1)n(2(n - 1) + 1) = \frac{1}{6}(n - 1)n(2n - 1).$$

The area is therefore

$$A = \frac{x^3}{n^3}\frac{1}{6}(n - 1)n(2n - 1),$$

$$= \frac{1}{3}x^3\left(1 - \frac{1}{n}\right)\left(1 - \frac{1}{2n}\right).$$

Again it is clear that if a larger number of strips are taken the accuracy improves and so taking the limit as $n \to \infty$ should give absolute accuracy.

$$A = \frac{1}{3}x^3.$$

This is the area from 0 to x and so the area from 0 to 5 is $\frac{1}{3}5^3$ and from 0 to 2 is $\frac{1}{3}2^3$. The area from 2 to 5 is therefore

$$\frac{1}{3}(5^3 - 2^3) = 39,$$

which is close to the original estimate of about 40.

The original equation was

$$y = x^2$$

and the area from 0 to x has been found to be

$$A = \frac{1}{3}x^3.$$

It is worth noting that x^2 is the derivative of $\frac{1}{3}x^3$, or, conversely, the area A is the anti-derivative of x^2. It is the graph of $y = x^2$ under which the area is being calculated.

5.3. The integral

At the end of the previous section a connection between the area under a curve and the anti-derivative of §5.1 was indicated. If such a connection is to be of any value it should be established on a more general basis. To attempt this the area under the graph of

$$y = f(x)$$

from $x = a$ to $x = b$ is considered and is shown in Figure 5.10. The area can be divided into strips and a typical strip can be represented as starting at x and being of width δx. The left-hand side of the strip will be of height $f(x)$ and the right-hand side will be of height $f(x + \delta x)$.

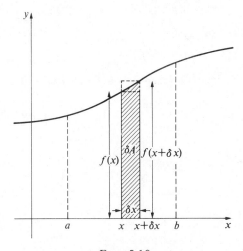

FIG. 5.10

The exact area, δA, of the strip will therefore lie between $f(x)\delta x$ and $f(x + \delta x)\delta x$. As δx decreases these two estimates converge towards each other with the actual area trapped between them. The total area A will be the sum of all such strips

$$A \approx \sum_{x=a}^{x=b} f(x)\delta x.$$

As the strip width decreases to zero the approximation becomes an equality and the symbols used change as they did in the case of the derivative. The δx becomes $\mathrm{d}x$ and the Σ, meaning summation, is replaced by an extended letter 'S' for summation.

$$\lim_{\delta x \to 0} \sum_{x=a}^{x=b} f(x)\delta x = \int_a^b f(x)\mathrm{d}x = A$$

This is known as a *definite integral* since it refers to a specific range of values of x and hence to a specific area. The numbers a and b are known as the *lower* and *upper limits of integration*, respectively. The function $f(x)$ inside the integral is known as the *integrand*.

Referring to Figure 5.10 it can be seen that if $f(x)$ is increasing over the range x to $x + \delta x$, then $f(x) < f(x + \delta x)$, and so

$$f(x)\delta x < \delta A < f(x + \delta x)\delta x.$$

If however $f(x)$ is decreasing,

$$f(x)\delta x > \delta A > f(x + \delta x)\delta x.$$

If the possibility of $f(x)$ remaining stationary is included in each case, we have

$$f(x) \leqslant \frac{\delta A}{\delta x} \leqslant f(x + \delta x)$$

or

$$f(x) \geqslant \frac{\delta A}{\delta x} \geqslant f(x + \delta x).$$

In either situation as the strip becomes narrower, $\delta x \to 0$ and $f(x + \delta x) \to f(x)$ with $\delta A/\delta x$ trapped between. In the limit therefore

$$\frac{\mathrm{d}A}{\mathrm{d}x} = f(x)$$

and so A is the anti-derivative of $f(x)$ (§5.1) and the connection between the area under a curve and the anti-derivative is established. Since $f(x)$ is a function of x the area A will also be a function of x and we may write

$$A = A(x) = F(x) + C$$

where $F(x)$ is the anti-derivative of $f(x)$. Because no specific area is referred to this is known as an *indefinite integral* and written

$$\int f(x)\mathrm{d}x = F(x) + C.$$

It will always contain an arbitrary constant the value of which cannot be determined until a starting point from which the area is measured is known.

Let us assume that a starting point from which the area is measured is $x = a$. If this is the case then

$$A(a) = 0$$

because this represents the area from $x = a$ to $x = a$ and hence no area at all. However

$$A(x) = F(x) + C$$

and so

$$A(a) = F(a) + C = 0,$$

giving

$$C = -F(a).$$

The general area function can therefore be written

$$A(x) = F(x) - F(a).$$

The specific area from $x = a$ to $x = b$ is then

$$A(b) = \int_{a}^{b} f(x)\mathrm{d}x = F(b) - F(a).$$

A shorthand intermediate step is often included so that we write

$$\int_{a}^{b} f(x)\mathrm{d}x = F(x)\Big|_{a}^{b} = F(b) - F(a).$$

This signifies that first of all the anti-derivative or indefinite integral is determined and is to be evaluated between the two specific limits a and b. The final step represents this evaluation. It is this formulation which is invariably used when an area is being calculated, that is, an *integration* is being performed. It is easy to see that there is no need to include the arbitrary constant, C, from the indefinite integral when a definite integral is being evaluated because it cancels out.

$$\int_a^b f(x)dx = (F(x) + C)\Big|_a^b = (F(b) + C) - (F(a) + C)$$

$$= F(b) - F(a).$$

The key to calculating any area or evaluating any definite integral is a knowledge of the anti-derivative or indefinite integral of the function involved. In §5.1 it was shown how a knowledge of derivatives can lead to the formation of a table of anti-derivatives which we now know are the same as indefinite integrals. A table of indefinite integrals is given in Appendix B.3. Extensive lists are published in some books of mathematical tables and formulae.

There are some general formulae involving integrals which arise as a direct consequence of the equivalent formulae for derivatives.

1. $$\int kf(x)dx = k \int f(x)dx = kF(x) + C$$

2. $$\int (f(x) \pm g(x))dx = \int f(x)dx \pm \int g(x)dx = F(x) \pm G(x) + C$$

3. A useful formula can be derived from the derivative of a product. If u and v are functions of x then rule 3, §4.5 shows that

$$\frac{d}{dx}(uv) = u\frac{dv}{dx} + v\frac{du}{dx}.$$

Hence,

$$\int \frac{d}{dx}(uv)dx = \int u\frac{du}{dx}dx + \int v\frac{dv}{dx}dx,$$

$$\int d(uv) = \int udv + \int vdu.$$

Therefore, $$\int udv = uv - \int vdu.$$

This is the *integration by parts formula*.

The ways in which these formulae or rules operate are best illustrated by means of examples.

Example. Find the area between $y = x^2 + 3x$ and the x-axis bounded by $x = 1$ and $x = 2$.

It is usually helpful to sketch the area concerned and this is done in Figure 5.11. There appears to be nothing to cause any complications.

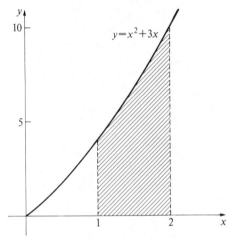

FIG. 5.11

The area is given by

$$\int_1^2 (x^2 + 3x)dx = \int_1^2 x^2\,dx + \int_1^2 3xdx \qquad \text{(Rule 2)}$$

$$= \int_1^2 x^2dx + 3\int_1^2 xdx \qquad \text{(Rule 1)}$$

$$= \frac{1}{3}x^3 \Big|_1^2 + 3\frac{x^2}{2}\Big|_1^2$$

$$= \frac{1}{3}(8 - 1) + \frac{3}{2}(4 - 1)$$

$$= \frac{41}{6}.$$

Example. The rate R (l/min) at which blood is being pumped through an artery at any instant in time t (min) is given by

$$R = \frac{b}{15}\{1 + \sin(2\pi bt)\}$$

where b is the heart rate (beats/min). Find the total volume of blood which is pumped through the artery in one hour at heart rates of 70 beats per minute and 120 beats per minute.

The total volume V is the area under the graph of R against t from $t = 0$ to $t = 60$. The rate R is never negative and so

$$V = \int_0^{60} R\,dt = \int_0^{60} \frac{b}{15}\left\{1 + \sin(2\pi bt)\right\} dt$$

$$= \frac{b}{15}\left\{t - \frac{1}{2\pi b}\cos(2\pi bt)\right\}\Big|_0^{60}$$

$$= \frac{b}{15}\left\{(60 - 0) - \frac{1}{2\pi b}(\cos(120\pi b) - \cos(0))\right\}$$

$$= 4b \text{ provided } 60b \text{ is an integer.}$$

Hence at 70 beats per minute $V = 280$ l and at 120 beats per minute $V = 480$ l. In these cases the trigonometric function has not contributed at all. It cancelled itself out between the upper and lower limits.

Example. Find $\int x \sin x \, dx$.

This is an indefinite integral and so a specific area cannot be drawn. The function itself cannot be integrated as it stands but can be simplified using integration by parts. Note that in this formula one term, u, is differentiated and the other, dv, is integrated. This gives a new integral $\int v\,du$ which may be simpler. In this particular case differentiating x eliminates it and so we choose $x = u$, leaving

$$\sin x\,dx = dv.$$

Therefore, $du = dx$ and $v = \int \sin x\,dx = -\cos x.$

Hence,

$$\int x \sin x \, dx = -x \cos x - \int (-\cos x) \, dx$$

$$= -x \cos x + \int \cos x \, dx$$

$$= -x \cos x + \sin x + C.$$

Note that any indefinite integral can be checked by differentiation which should give the original integrand.

$$\frac{d}{dx}(-x \cos x + \sin x + C) = -\frac{d}{dx}(x \cos x) + \cos x$$

$$= -(-x \sin x + \cos x) + \cos x$$

$$= x \sin x \qquad \text{Checks!}$$

Example. Find the area between $y = \cos x$ and the x-axis from $x = 0$ to $x = \pi$.

A sketch of the area (Figure 5.12) shows that part of it is below the x-axis which is a situation not encountered before. This may have unforeseen implications. A direct integration gives the area as

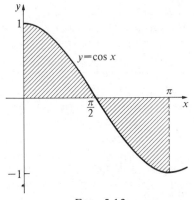

FIG. 5.12

$$\int_0^\pi \cos x \, dx = \sin x \, \Big|_0^\pi = \sin \pi - \sin 0 = 0.$$

The calculated area is zero but clearly from the diagram an area does exist. Since the total area has two parts, one above and one below the x-axis, we can try calculating each independently. The upper area is

$$\int_0^{\frac{\pi}{2}} \cos x \, dx = \sin x \, \Big|_0^{\frac{\pi}{2}} = \sin \frac{\pi}{2} - \sin 0 = 1.$$

The lower area is given by

$$\int_{\frac{\pi}{2}}^\pi \cos x \, dx = \sin x \, \Big|_{\frac{\pi}{2}}^\pi = \sin \pi - \sin \frac{\pi}{2} = -1.$$

This second area is negative and when added to the first area exactly cancels it out. If the minus sign is neglected the total area is seen to be 2 units but some explanation of this sign is required.

In practical situations most graphs only exist in the first quadrant which is why, when a section of curve is being drawn, it tends to be drawn in this quadrant. This has been the case in the diagrams associated with the theory of this and previous sections. A curve has been drawn and an area in the first quadrant bounded by the curve, the x-axis and two x values have been considered. In general situations these have been $x = a$ and $x = b$, where, in diagrams, $b > a$. The thin strips have had a width δx where δx was some small interval measured from left to right, that is, in a positive x-direction, so that $\delta x > 0$. For a curve in the first quadrant the height of the strip is the value the function takes at the top of the strip and is positive. The thin strip therefore has a positive area since it consists of the product of $f(x)$ and δx, both of which are positive. Summing such strips to give the complete area gives a positive result because each strip in the first quadrant has a positive area.

Now consider the situation for a section of curve below the x-axis as shown in Figure 5.13. In this case the height of the strip, still defined to be the value the function takes, is negative. The interval δx is positive still and so the strip area, $\delta A = f(x)\delta x$, is negative. This explains the results obtained in the last example. The summation of such strips produces a negative area which may partially or completely cancel out some positive area elsewhere.

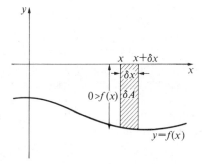

F IG . 5.13

It is the sign of the product $f(x)\delta x$ which determines whether an area will be positive or negative. For an area in the first quadrant reversing the interval a to b will make the area negative. In considering strips from b to a rather than from a to b, where $b > a$, the width δx must be negative because $x = b$ is being reduced in steps of δx down to $x = a$. The change in sign is clear from the definite integral,

$$\int_a^b f(x)\mathrm{d}x = F(b) - F(a)$$

where the range of integration is from a to b. For a range b to a the limits are interchanged to give

$$\int_b^a f(x)\mathrm{d}x = F(a) - F(b) = -\int_a^b f(x)\mathrm{d}x.$$

Hence reversing the limits will always change the sign of the integral. It is conventional to have the range of integration running from left to right along the horizontal axis so that the upper limit of integration is larger than the lower limit. If a situation arises in which this is not the case the limits can be interchanged provided a corresponding change is made to the sign in front of the integral.

When the area under a curve was first considered we looked at the area bounded by the curve, the two coordinate axes, and some vertical line. This was generalized to cover the area bounded by the curve, the horizontal axis, and the two vertical lines $x = a$ and $x = b$ giving the definite integral

$$A = \int_a^b f(x)\mathrm{d}x.$$

There is no reason why one horizontal bound should be the horizontal axis. The area between the curve $y = f(x)$ and the horizontal line $y = c$ is still an area which it should be possible to calculate. One way of doing this is to find the area between $y = f(x)$ and the x-axis and subtract the area between $y = c$ and the x-axis. Such an area is shown in Figure 5.14 and can be found from

$$A = \int_a^b f(x)\mathrm{d}x - \int_a^b c\,\mathrm{d}x$$

$$= \int_a^b f(x)\mathrm{d}x - c(b - a).$$

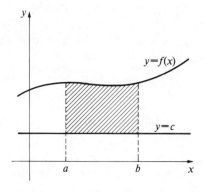

FIG. 5.14

This last term is the area of the rectangle between $y = c$ and the x-axis, $y = 0$. In the same way the area between two curves $y = f(x)$ and $y = g(x)$ (Figure 5.15) can be found by subtraction. The area is

$$A = \int_a^b f(x)\mathrm{d}x - \int_a^b g(x)\mathrm{d}x$$

$$= \int_a^b \{f(x) - g(x)\}\mathrm{d}x.$$

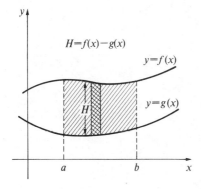

F IG . 5.15

This last expression shows that we are calculating the area of thin strips of height $f(x) - g(x)$ and width δx. The advantage in using this form is that some cancellation may take place between $f(x)$ and $g(x)$ on subtraction and so save on integration. When dealing with the area between two functions it is most important to be aware of what is happening. If they cross each other it may not be clear just what area is being calculated and so sketches can be of great value.

Example. Find the area bounded by $y = x^2 - 4x + 5$ and $y = x + 1$.

Since no limits on x are given the line must cross the parabola and to find where this occurs we solve

$$x^2 - 4x + 5 = x + 1.$$

Therefore, $x^2 - 5x + 4 = (x - 1)(x - 4) = 0.$

Hence, the line crosses the parabola at $(1, 2)$ and $(4, 5)$. A diàgram of the situation is shown in Figure 5.16. The higher of the two is the line $y = x + 1$ and so to obtain a positive area the equation of the parabola is subtracted from this.

$$f(x) = x + 1,$$
$$g(x) = x^2 - 4x + 5.$$

The limits of integration are determined by the points of intersection and so they are $x = 1$ and $x = 4$. The area is

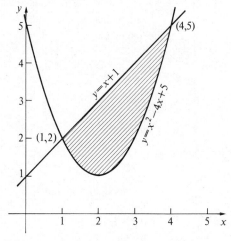

$$A = \int_a^b \left\{ f(x) - g(x) \right\} dx = \int_1^4 \left\{ (x + 1) - (x^2 - 4x + 5) \right\} dx$$

$$= \int_1^4 \left\{ -x^2 + 5x - 4 \right\} dx = \left\{ -\frac{1}{3}x^3 + \frac{5}{2}x^2 - 4x \right\} \Big|_1^4$$

$$= \left(-\frac{1}{3}4^3 + \frac{5}{2}4^2 - 4^2 \right) - \left(-\frac{1}{3} + \frac{5}{2} - 4 \right)$$

$$= 4\tfrac{1}{2}.$$

A feature related to the area below the graph of $y = f(x)$ is the *mean value* of this function over a specified interval. To see how this is related to the more conventional idea of the mean of a set of numbers defined by

$$\bar{x} = \frac{1}{n} \sum_{i=1}^n x_i,$$

the area is split into a series of strips in the now familiar way. The mean

height of all such strips is the sum of all the heights divided by the
number of strips. The height of each strip depends upon the value of
the function at that point as shown in Figure 5.17. Assume that the
interval between $x = a$ and $x = b$ is divided up into n equal strips each

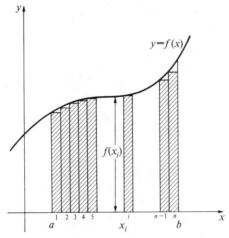

F IG . 5.17

of width $(b - a)/n$. The height of the strip at the point $x = x_i$ is $f(x_i)$
and there will be n such points over the interval a to b. The mean
height $\bar{H}$ is therefore given by

$$\bar{H} = \frac{1}{n} \sum_{i=1}^{n} f(x_i) = \sum_{i=1}^{n} \frac{f(x_i)}{n}.$$

As n increases the number of strips increases and their width decreases.
More and more height samples of the function are being taken and so
$\bar{H}$ is beginning to represent not just individual heights but a measure of
all values taken by the function. If we denote the strip width by δx
then multiplying the right-hand side by $\delta x/\delta x$ leaves it unaltered and
we have

$$\bar{H} = \frac{1}{n\delta x} \sum_{i=1}^{n} f(x_i)\delta x$$

$$= \frac{1}{\dfrac{(b-a)}{n}} \sum_{i=1}^{n} f(x_i)\delta x$$

$$= \frac{1}{b-a} \sum_{i=1}^{n} f(x_i)\delta x$$

since the strip width is also $(b-a)/n$. As n tends to infinity the interval a to b is sampled at all values of x within it and the summation becomes an integral from $x = a$ to $x = b$. Since it is the value of the function that is being sampled at all points within this range the integral represents the mean value of the function over this range. Hence,

$$\overline{f(x)} = \frac{1}{b-a} \int_a^b f(x)\mathrm{d}x.$$

Geometrically this shows that the area below the curve from $x = a$ to $x = b$ is the same as that of a rectangle of base $(b-a)$ and height $\overline{f(x)}$.

Example. Find the mean value of $\sin x$ in the interval $0 \leqslant x \leqslant \pi$.

The mean is

$$\frac{1}{\pi} \int_0^\pi \sin x \mathrm{d}x = \frac{1}{\pi} (-\cos x) \Big|_0^\pi = \frac{2}{\pi}.$$

Example. The gross photosynthetic rate R of a plant is related to the rate at which light energy E falls upon it by

$$R = \frac{1}{4.0 + \dfrac{0.1}{E}}.$$

Find the mean value of R for a range of light energy rates from 0 to 0.1 $\mathrm{Wm^{-2}}$.

The required mean value is given by

$$\bar{R} = \frac{1}{0.1 - 0} \int_0^{0.1} R\,dE$$

$$= 10 \int_0^{0.1} \frac{1}{4.0 + \dfrac{0.1}{E}}\,dE = 10 \int_0^{0.1} \frac{E}{4.0E + 0.1}\,dE$$

$$= 10 \int_0^{0.1} \left(\frac{1}{4} - \frac{\dfrac{1}{40}}{4E + \dfrac{1}{10}} \right)\,dE$$

$$= 10 \left(\frac{1}{4}E - \frac{1}{160} \ln \left| 4E + \frac{1}{10} \right| \right) \Big|_0^{0.1}$$

$$= 10 \left(\frac{1}{4}(0.1) - \frac{1}{160} \ln \left| \frac{0.4 + 0.1}{0.1} \right| \right)$$

$$= \frac{1}{4} \left(1 - \frac{1}{4} \ln 5 \right) = 0.149.$$

Exercises

1. Find the areas bounded by the given curves, the x-axis ($y = 0$) and the given vertical lines.

(a) $y = x^2 + 3x$ between $x = 2$ and $x = 3$;

(b) $y = 2x - 7$ between $x = 4$ and $x = 10$;

(c) $y = -x^2 + 7x - 10$ between $x = 2$ and $x = 5$;

(d) $y = e^x$ between $x = -1$ and $x = 2$;

(e) $y = \cos x$ between $x = -\frac{1}{4}\pi$ and $x = \frac{1}{2}\pi$.

2. Find the indefinite integrals of the following functions.

(a) $x^2 - 4x + 3$;   (b) $4x - \sin x$;   (c) $\sqrt{x}$;

(d) $x \cos x$;   (e) $x + \dfrac{1}{x}$;   (f) $x^2 e^x$.

3. By considering a graph of $y = x^2 - 6x + 8$ for values of x between 0 and 4 determine which areas between this curve and the x–axis are positive and which are negative. Find the area in each case.

4. Find the area between $y = 4x - x^2$ and $y = x^2 - 2x + 4$.

5. The population P_L of lynxes in a certain area of Canada is given by

$$P_L = 40\ 000 + 35\ 000 \sin\left(\frac{2}{11}\pi t\right).$$

Find the mean value of P_L over one period.

5.4. The integral of $1/x$

In general

$$\int x^\alpha \mathrm{d}x = \frac{x^{\alpha+1}}{\alpha + 1} + C$$

but $\alpha = -1$ is excluded. It is clear that if it were included the result above would be impossible to interpret. This is because $\alpha + 1$ would be $-1 + 1 = 0$ and we would have

$$\int \frac{1}{x} \mathrm{d}x = \frac{x^0}{0} + C.$$

However a graph of $y = 1/x$ can be drawn and the area below it from $x = a$ to $x = b$ can certainly be defined. Such an area is shown in Figure 5.18. If values of a and b are specified an accurate graph can be drawn and a good approximation to the area found. It seems therefore that it should be possible to give a meaning to the definite integral

$$\int_a^b \frac{1}{x} \mathrm{d}x$$

which represents a particular area. To make the situation as simple as possible a particular value for the lower limit a could be chosen. A natural choice would be $a = 0$ but from the graph of $y = 1/x$ it can be seen that the curve goes off to infinity here and so y is undefined. It appears wise to exclude $a = 0$ and so we could try $a = 1$. This gives a well-defined area as shown in Figure 5.19, where a variable upper

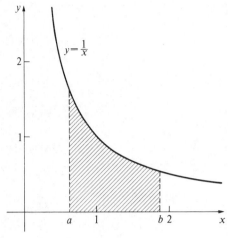

FIG. 5.18

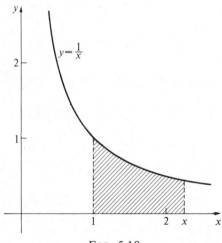

FIG. 5.19

bound of x has been taken. The area will therefore be a function of x and will be denoted by $L(x)$.

Therefore,
$$L(x) = \int_1^x \frac{\mathrm{d}x}{x}.$$

It is possible from this alone to derive a number of properties of the function $L(x)$.

1.
$$L(1) = \int_1^1 \frac{dx}{x} = 0.$$

2.
$$L(ab) = \int_1^{ab} \frac{dx}{x} = \int_1^a \frac{dx}{x} + \int_a^{ab} \frac{dx}{x}.$$

In the second integral put $x = au$ and change the limits accordingly. dx becomes adu since $dx/du = a$.

Therefore, $$L(ab) = \int_1^a \frac{dx}{x} + \int_1^b \frac{du}{u} = L(a) + L(b).$$

3.   From 1 and 2
$$L(1) = L\left(b \times \frac{1}{b}\right) = L(b) + L\left(\frac{1}{b}\right) = 0.$$

Therefore,
$$L\left(\frac{1}{b}\right) = -L(b).$$

4.   From 2 and 3
$$L\left(\frac{a}{b}\right) = L\left(a \times \frac{1}{b}\right) = L(a) + L\left(\frac{1}{b}\right) = L(a) - L(b).$$

These properties are characteristic of a logarithmic function and so we may write
$$L(x) = \int_1^x \frac{dx}{x} = \log_a x.$$

For any logarithm $\log_a a = 1$ and in this case it means that
$$\int_1^a \frac{dx}{x} = 1.$$

Some relatively simple approximations to the value of a can be made. The area for different values of a can be estimated and a range of values found which come close to satisfying the condition of unit area. A reasonably good approximation can be made to the curve with a series of chords and the trapezoidal area below each of these chords calculated. Figure 5.20 shows the situation with a horizontal step interval

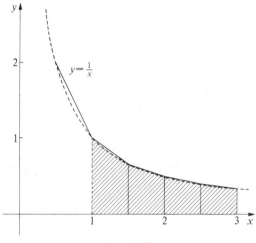

F IG . 5.20

of $\frac{1}{2}$. The area of a trapezium is its average height multiplied by its width and so the following table is produced.

x	1.0	1.5	2.0	2.5	3.0
y	1	$\frac{2}{3}$	$\frac{1}{2}$	$\frac{2}{5}$	$\frac{1}{3}$
Area of trapezium	0.4167	0.2917	0.2250	0.1833	
Cumulative area	0.4167	0.7083	0.9333	1.1167	

This shows that the appropriate base lies between 2.5 and 3.0 and linear interpolation (§3.7) gives a value of about 2.7. Obviously a smaller step interval would give a more accurate approximation. It can be shown that the base is the number e $= 2.71828 \ldots$ and so the function $L(x)$ is the *natural logarithm* $\ln x$.

$$\int_1^x \frac{\mathrm{d}x}{x} = \ln x.$$

The corresponding indefinite integral is

$$\int \frac{dx}{x} = \ln x + C.$$

Differentiating both sides of this equation and recalling that differentiation and integration are inverse operations with respect to each other leaves the left-hand side as $1/x$ and the right-hand side as $(d/dx)\ln x$. Hence,

$$\frac{d}{dx} \ln x = \frac{1}{x}$$

which is the key result quoted in §4.5.

The natural logarithm and hence its base e have arisen in a natural way from a consideration of the area below $y = 1/x$. It is perhaps in some way analogous to the manner in which π arises naturally when investigating the area of a circle. Both e and π are *transcendental* numbers. They cannot be solutions of purely algebraic equations.

5.5. Integration techniques

Integration is in general a more difficult process than differentiation. Whereas derivatives of most functions can be found without too much trouble the same cannot be said of integrals. Although a graph of a particular function can be drawn and the area represented by an integral shown on it there is no guarantee that the integration can be performed by analytic methods. An integral in the form of a function of the independent variable may not even exist. In this section a number of fairly standard approaches are outlined which if applied systematically to an integral will improve the chances of success. If these techniques fail then the numerical methods given in the following section will have to be employed.

Because integration is the reverse of differentiation we can expect to be able to evaluate a fair number of the simpler looking integrals by comparing the integrands with known derivatives. For every derivative listed a corresponding integral can be listed and a number of these *standard forms* are given in Appendix B.3. If an integral can be converted into one of these forms it can be evaluated. There will however be situations in which such a conversion is not obvious and some intuition, experience, or even guesswork has to be used. The techniques which can be employed are best illustrated by application to a number of examples with comments being made where appropriate.

Example. Find

$$\int \frac{1}{\sqrt{(1+x)}}.$$

It may be recalled that

$$\frac{d}{dx}\sqrt{x} = \frac{1}{2\sqrt{x}}$$

and so we might expect the integral to look like $\sqrt{(1+x)}$. Such a guess can be checked for validity by differentiation.

$$\frac{d}{dx}\sqrt{(1+x)} = \frac{1}{2}(1+x)^{\frac{1}{2}} = \frac{1}{2\sqrt{(1+x)}}.$$

Hence,

$$\int \frac{dx}{\sqrt{(1+x)}} = 2\sqrt{(1+x)} + C.$$

Alternatively a substitution could have been chosen to try and simplify the integral. Eliminating the square root might help so we can try $1 + x = u^2$. The dx in the integral must also be substituted and differentiation shows that $dx = 2u\,du$. The integral can therefore be converted to

$$\int \frac{dx}{\sqrt{(1+x)}} = \frac{2u\,du}{u} = 2\int du = 2u + C.$$

Substituting back to put the answer into the original variable shows that

$$\int \frac{dx}{\sqrt{(1+x)}} = 2\sqrt{(1+x)} + C$$

as before.

Example. Find

$$\int \frac{\cos x}{\sqrt{\sin x}}\,dx.$$

This looks particularly awkward but eliminate the square root as in the previous example. Put $\sin x = u^2$, $\cos x\,dx = 2u\,du$.

Therefore,

$$\int \frac{\cos x dx}{\sqrt{\sin x}} = \int \frac{2u du}{u} = 2\int du = 2u + C = 2\sqrt{\sin x} + C.$$

Example. Find

$$\int x\sqrt{(1 + x^2)} dx.$$

Try $1 + x^2 = u^2$, $2x dx = 2u du$,

$$\int x\sqrt{(1 + x^2)} dx = \int u^2 du = \frac{1}{3}u^3 + C = \frac{1}{3}(1 + x^2)^{\frac{3}{2}} + C.$$

Example. Find

$$\int \frac{x}{x^2 + 4} dx.$$

This is a particular example of the more general case

$$\int \frac{f'(x)}{f(x)} dx = \ln(f(x)) + C,$$

which is a direct consequence of the derivative of $\ln(f(x))$ using the chain rule. The integral can be found by writing

$$\int \frac{x}{x^2 + 4} dx = \frac{1}{2}\int \frac{2x dx}{x^2 + 4} = \frac{1}{2}\ln(x^2 + 4) + C.$$

Two not so obvious results arising in a similar manner are

$$\tan x dx = \int \frac{\sin x}{\cos x} dx = -\int \frac{-\sin x}{\cos x} dx = -\ln(\cos x) + C$$

and

$$\int \cot x dx = \ln \sin x + C.$$

The integral of the product of a function and its derivative can also be found

$$\int f(x)f'(x)dx = \frac{1}{2}\{f(x)\}^2 + C$$

e.g. $\int \tan x \sec^2 x dx = \frac{1}{2}\tan^2 x + C.$

If a simple substitution does not appear to work something more sophisticated may have to be tried. In situations where $a^2 - x^2$ or $a^2 + x^2$ occur the substitutions $x = a \sin \theta$ and $x = a \tan \theta$ respectively might be of value. The identities $1 - \sin^2 \theta = \cos^2 \theta$ and $1 + \tan^2 \theta = \sec^2 \theta$ are then usually employed. Once trigonometric functions have been introduced various identities may have to be used. Some of these are given in Appendix B.1.

Example. Find

$$\int \sqrt{(4 - x^2)}dx.$$

Try $4 - x^2 = u^2$, $-xdx = udu$. There is no extra x to go with the dx and so not much progress can be expected. Try instead $x = 2 \sin \theta$, $dx = 2 \cos \theta d\theta$.

$$\int \sqrt{(4 - x^2)}dx = \int \sqrt{(4 - 4 \sin^2 \theta)} \, 2 \cos \theta d\theta = \int 4 \cos^2 \theta d\theta$$

$$= 2 \int 2 \cos^2 \theta d\theta = 2 \int (\cos 2\theta + 1)d\theta$$

$$= 2\left(\frac{\sin 2\theta}{2} + \theta\right) + C.$$

To get this far we have used $1 - \sin^2 \theta = \cos^2 \theta$ and $\cos 2\theta = 2 \cos^2 \theta - 1$. We must now substitute back for the original variable x. Using $\sin 2\theta = 2 \sin \theta \cos \theta$ enables us to put $\sin \theta = x/2$ but $\cos \theta$ must be calculated in terms of x from this. Now $\cos \theta = \sqrt{(1 - \sin^2 \theta)} = \sqrt{(1 - (x/2)^2)}$, and so we have

$$\int \sqrt{(4 - x^2)}dx = 2(\sin \theta \cos \theta + \theta) + C$$

$$= 2\left\{\frac{x}{2}\sqrt{\left(1 - \left(\frac{x}{2}\right)^2\right)} + \theta\right\} + C.$$

Now θ is an angle and from the original substitution it is the angle such that $\sin \theta = x/2$. Since x is not known θ cannot be evaluated. In this situation an *inverse function* notation is used. We write

$$\theta = \sin^{-1}\left(\frac{x}{2}\right)$$

which means simply that θ is the angle whose sine is $x/2$. The complete integral is therefore

$$\int \sqrt{(4-x^2)}\mathrm{d}x = \frac{1}{2}x\sqrt{(4-x^2)} + 2 \sin^{-1}\left(\frac{x}{2}\right) + C.$$

This deceptively simple integral has proved quite difficult to find. Two integrals yielding inverse functions alone are

$$\int \frac{\mathrm{d}x}{\sqrt{(a^2 - x^2)}} = \sin^{-1}\left(\frac{x}{a}\right) + C$$

and

$$\int \frac{\mathrm{d}x}{x^2 + a^2} = \frac{1}{a}\tan^{-1}\left(\frac{x}{a}\right).$$

The first can be proved by putting $x = a \sin \theta$ and the second by $x = a \tan \theta$. Care should be taken when using inverse functions to avoid interpreting the $^{-1}$ notation as an exponent.

The integration by parts formula of §5.3

$$\int u\mathrm{d}v = uv - \int v\mathrm{d}u$$

can be applied in situations where part of the integrand becomes simpler on either differentiation or integration.

Example. Find $\int \ln x\mathrm{d}x$.

The logarithmic function becomes a simple rational function on differentiation and so we can try $u = \ln x$, $\mathrm{d}v = \mathrm{d}x$ giving $\mathrm{d}u = (1/x)\mathrm{d}x$ and $v = x$. The formula gives

$$\int \ln x\mathrm{d}x = x \ln x - \int x\frac{1}{x}\mathrm{d}x$$

$$= x \ln x - x + C.$$

In certain circumstances two or more applications may be required.

Example. Find $\int x^2 e^x dx$.

The exponential is unaltered by either differentiation or integration. The term x^2 is however reduced to $2x$ after one differentiation and to 2 after a second. We can therefore try $u = x^2$ and $dv = e^x dx$ giving $du = 2x dx$ and $v = e^x$.

$$\int x^2 e^x dx = x^2 e^x - \int 2x e^x dx$$

$$= x^2 e^x - 2(x e^x - \int e^x dx)$$

$$= x^2 e^x - 2(x e^x - e^x) + C$$

$$= (x^2 - 2x + 2)e^x + C$$

where in line two a second application of the formula was employed.

Example. Find $\int e^{ax} \sin bx dx$.

In this case the exponential remains an exponential and the sine alternates between cosine and sine on repeated differentiation or integration. Integration by parts can however be of value. Put $u = e^{ax}$ and $\sin bx dx = dv$ so that $du = a e^{ax} dx$ and $v = -(1/b) \cos bx$.

Therefore,

$$\int e^{ax} \sin bx dx = -\frac{1}{b} e^{ax} \cos bx + \frac{a}{b} \int e^{ax} \cos bx dx.$$

Now use integration by parts again by putting $u = e^{ax}$ and $du = \cos bx dx$ giving $du = a e^{ax} dx$ and $v = (1/b) \sin bx$. Care must be taken not to integrate the e^{ax} or the result will just give back the original integral and nothing else.

$$\int e^{ax} \sin bx dx = -\frac{1}{b} e^{ax} \cos bx + \frac{a}{b} \left\{ \frac{1}{b} e^{ax} \sin bx - \frac{a}{b} \int e^{ax} \sin bx dx \right\}$$

Hence

$$\left(1 + \frac{a^2}{b^2}\right) \int e^{ax} \sin bx dx = \frac{e^{ax}}{b}\left(\frac{a}{b}\sin bx - \cos bx\right) + C.$$

Therefore,

$$\int e^{ax} \sin bx dx = \frac{e^{ax}}{a^2 + b^2}\, (a \sin bx - b \cos bx) + C.$$

Example. A decaying population of size P which exhibits annual oscillations may be represented by

$$P = 1500\{4 + \sin(2\pi t)\}e^{-\frac{t}{10}}$$

where t is the time in years. Find the mean size of the population over the first five years.

The mean size $\bar{P}$ of the population over the first five years is given by the integral

$$\bar{P} = \frac{1}{5}\int_0^5 P dt = \frac{1}{5}\int_0^5 1500\{4 + \sin(2\pi t)\}e^{-\frac{t}{10}}dt$$

$$= 1200\int_0^5 e^{-\frac{t}{10}}\, dt + 300\int_0^5 e^{-\frac{t}{10}}\sin(2\pi t)dt.$$

The second of these integrals is of the form evaluated above and so we may write

$$\bar{P} = 1200(-10)e^{-\frac{t}{10}}\Big|_0^5 + 300\,\frac{e^{-\frac{t}{10}}}{\frac{1}{100} + 4\pi^2}\left\{-\frac{1}{10}\sin(2\pi t) - 2\pi \cos(2\pi t)\right\}\Big|_0^5$$

$$= 12000(1 - e^{-\frac{1}{2}}) + \frac{30000}{400\pi^2 + 1}\,2\pi(1 - e^{-\frac{1}{2}})$$

$$= 12000\left(1 - \frac{1}{\sqrt{e}}\right)\left(1 + \frac{5\pi}{400\pi^2 + 1}\right)$$

$$= 4740.$$

It is interesting to note that the oscillations have very little effect on this mean value. Their contribution is contained in the $5\pi/(400\pi^2 + 1)$ of the last expression. This is 0.00398 compared with the 1 which comes from the non-oscillating term. If the population had been simply decaying without oscillation then

$$P = 6000e^{-\frac{t}{10}}$$

and

$$\bar{P} = \frac{1}{5} \int_0^5 6000e^{-\frac{t}{10}} dt = 1200(-10)e^{-\frac{t}{10}} \Big|_0^5$$

$$= 12000 \left(1 - \frac{1}{\sqrt{e}}\right) = 4722.$$

Rational functions (§2.5) occur quite frequently and their integrals are often required. In §5.1 a differential equation for bounded population growth was derived which had the form

$$\frac{dP}{dt} = k(P_M - P).$$

This can be rearranged to give

$$\frac{dP}{P_M - P} = k dt$$

and integrated,

$$\int \frac{dP}{P_M - P} = \int k dt.$$

The first integral is of a simple rational function and takes the form of a logarithm

$$\int \frac{dP}{P_M - P} = -\int \frac{-dP}{P_M - P} = -\ln(P_M - P) + C_1.$$

The second integral is straightforward,

$$\int k dt = kt + C_2.$$

Hence,

$$-\ln(P_M - P) = kt + C.$$

Note that only one arbitrary constant is needed since the sum or difference of arbitrary constants is still an arbitrary constant.

We might require the population P as a function of time and so some rearrangement is required.

$$\ln(P_M - P) = -kt - C.$$

Hence, $$P_M - P = e^{-kt-C}.$$

Therefore, $$P = P_M - e^{-kt-C} = P_M - Ae^{-kt}$$

where $A = e^{-C}$ is an alternative arbitrary constant. This constant can only be determined if we have some further information, such as the population size at some time.

If a rational function is to be integrated and the degree of the numerator is higher than or equal to the degree of the denominator then the denominator should be divided into the numerator. The result will be a polynomial together with another rational function in which the numerator is of lower degree than the denominator.

Example. Find

$$\int \frac{x^2 + 2x + 2}{x + 3} dx.$$

Divide $x + 3$ into $x^2 + 2x + 2$ by long division.

$$
\begin{array}{r}
x + 3 \,) \overline{x^2 + 2x + 2} \,(x - 1 \\
\underline{x^2 + 3x} \\
-\ x + 2 \\
\underline{-\ x - 3} \\
5\,.
\end{array}
$$

Therefore, $$\frac{x^2 + 2x + 2}{x + 3} = x - 1 + \frac{5}{x + 3}.$$

This can be integrated with no problems.

$$\int \frac{x^2 + 2x + 2}{x + 3}\,dx = \int \left(x - 1 + \frac{5}{x + 3} \right) dx$$

$$= \frac{1}{2}x^2 - x + 5\int \frac{dx}{x + 3}$$

$$= \frac{1}{2}x^2 - x + 5\ln(x + 3) + C.$$

For a rational function in which the numerator is of lower degree than the denominator the factors, if any, of the denominator should be sought. If factors can be found the rational function can be split into a number of simpler *partial fractions*.

If these factors are *linear* and *not repeated* the rational function can be split into the sum of partial fractions each consisting of a constant in the numerator and one of the linear factors in the denominator. The constants can be determined by cross-multiplication and comparing the coefficients of each power of the variable on each side of the equation.

Example. Find the partial fractions of

$$\frac{4x + 7}{(x - 2)(2x + 1)}.$$

This can be written

$$\frac{4x + 7}{(x - 2)(2x + 1)} \equiv \frac{A}{x - 2} + \frac{B}{2x + 1}$$

since all the factors are linear and not repeated. Cross-multiplication gives

$$4x + 7 = A(2x + 1) + B(x - 2)$$

$$= (2A + B)x + (A - 2B).$$

Hence, $2A + B = 4$ and $A - 2B = 7$ giving $A = 3$ and $B = -2$.

Therefore, $\dfrac{4x + 7}{(x - 2)(2x + 1)} \equiv \dfrac{3}{x - 2} - \dfrac{2}{2x + 1}.$

The right-hand side is very much easier to integrate than the left.

If the factors are *linear* and one or more is *repeated*, then a number of separate partial fractions are required for the repeated factor. Each consists of a constant numerator but the denominators have the linear factor, the linear factor squared, etc., up to and including the degree to which it is repeated.

Example. Find the partial fractions of

$$\frac{4x}{(x-1)(x+1)^2}.$$

We write

$$\frac{4x}{(x-1)(x+1)^2} \equiv \frac{A}{x-1} + \frac{B}{x+1} + \frac{C}{(x+1)^2}$$

and find A, B, and C as before. $A = 1, B = -1, C = 2$.

If any of the factors are *quadratic* the numerator is a linear function.

Example. Find the partial fractions of

$$\frac{5x-7}{(x+3)(x^2+2)}.$$

This should be written as

$$\frac{5x-7}{(x+3)(x^2+2)} \equiv \frac{A}{x+3} + \frac{Bx+C}{x^2+2}$$

which gives $A = -2, B = 2$, and $C = -1$.

Example. Find

$$\int \frac{5x-7}{(x+3)(x^2+2)}\,dx.$$

Using partial fractions we have

$$\int \frac{5x-7}{(x+3)(x^2+2)}\,dx = -2\int \frac{dx}{x+3} + \int \frac{2x-1}{x^2+2}\,dx$$

$$= -2\int \frac{dx}{x+3} + \int \frac{2x}{x^2+2}\,dx - \int \frac{dx}{x^2+2}$$

$$= -2\ln(x+3) + \ln(x^2+2) - \frac{1}{\sqrt{2}}\tan^{-1}\left(\frac{x}{\sqrt{2}}\right) + C.$$

Whenever an integral like

$$\int \frac{2x-1}{x^2+2}\,dx$$

occurs the possibility of making the numerator into the derivative of the denominator should be investigated. This can invariably be accomplished leaving a second term with a constant numerator and the same denominator. The first term yields a logarithmic function after integration and the second may give an inverse tangent function.

Before concluding the discussion of partial fractions a quick way of obtaining the numerator constant for linear unrepeated factors should be mentioned. It is known as the 'cover up' method and can be illustrated by reconsidering the first partial fraction example. We have

$$\frac{4x+7}{(x-2)(2x+1)} \equiv \frac{A}{x-2} + \frac{B}{2x+1}.$$

To find A we choose the value of x which makes the denominator $x-2$ vanish. This is $x=2$. The factor $(x-2)$ is then 'covered up' on the left-hand side and $x=2$ substituted into what is left. The value obtained is the required value of A.

$$\frac{4x+7}{(XXX)(2x+1)} = \frac{4 \times 2 + 7}{(XXX)(2 \times 2 + 1)} = \frac{15}{(XXX)(5)}$$

and so $A=3$. Repeating the process for B requires $x=-\frac{1}{2}$ which gives

$$B = \frac{4 \times (-\frac{1}{2}) + 7}{(-\frac{1}{2} - 2)} = \frac{5}{(-\frac{5}{2})} = -2.$$

Example. In a unimolecular reaction between two compounds which produces a third the rate at which the molar concentration of the product is changing is given by

$$\frac{dz}{dt} = k(x_0 - z)(y_0 - z)$$

where k is a reaction constant and x_0 and y_0 are the initial molar concentrations of the original two compounds. If $x_0 = 3$ mol/l, $y_0 = 2$ mol/l, and $k = 0.0125$ find an expression for the molar concentration of the product, z, as a function of time t. Find also the value towards which z tends as t increases.

The rate at which z is changing is given by

$$\frac{dz}{dt} = 0.0125(3 - z)(2 - z)$$

which can be rearranged to give

$$\frac{dz}{(3 - z)(2 - z)} = 0.0125 dt.$$

At $t = 0$ the molar concentration of the product $z = 0$ and at an arbitrary time t this concentration is z say. We may therefore write

$$\int_0^z \frac{dz}{(3 - z)(2 - z)} = 0.0125 \int_0^t dt.$$

The left hand side is a rational function which can be split into partial fractions using the 'cover up' method to give

$$\int_0^z \left(\frac{1}{2 - z} - \frac{1}{3 - z} \right) dz = 0.0125 \int_0^t dt.$$

Hence

$$\ln\left(\frac{3 - z}{2 - z} \right) \bigg|_0^z = 0.0125 t \bigg|_0^t$$

and so

$$\ln\left(\frac{3 - z}{2 - z} \right) - \ln\left(\frac{3}{2} \right) = 0.0125 t.$$

Therefore

$$\frac{2}{3} \times \frac{3 - z}{2 - z} = e^{0.0125 t},$$

$$6 - 2z = (6 - 3z)e^{0.0125t}$$

giving

$$z = \frac{6(1 - e^{-0.0125t})}{3 - 2e^{-0.0125t}}.$$

As time increases the negative exponential tends towards zero and so z tends to 2.

The only way of becoming proficient at integration is by gaining considerable experience. There are few general rules which can be applied and it is by working through examples that insight can be obtained into the techniques most likely to bring success. There are books containing tables of large numbers of integrals of various kinds which can be of considerable use. Two such books are *Tables of Integrals and Other Mathematical Data* by H. B. Dwight (Collier-Macmillan) and *Mathematical Handbook of Formulae and Tables* by M. R. Spiegel (Schaum Outline Series).

Exercises

1. Find the following indefinite integrals.

(a) $\int \dfrac{dx}{\sqrt{(3 + 2x)}}$;   (b) $\int \sqrt{(4x - 1)}dx$;   (c) $\int \dfrac{dx}{(3x - 2)^2}$;

(d) $\int x\sqrt{(x^2 - 1)}dx$;   (e) $\int \sin 2x dx$;   (f) $\int \sin^2 x \cos x dx$;

(g) $\int \cos x \sqrt{(\sin x)}dx$;   (h) $\int \dfrac{dx}{2x}$;   (i) $\int \dfrac{dx}{2x + 7}$;

(j) $\int \dfrac{x + 1}{x^2 + 2x - 5}dx$;   (k) $\int \dfrac{\sin x}{\cos x}dx$;   (l) $\int \tan x dx$;

(m) $\int \sqrt{(9 - 4x^2)}dx$;   (n) $\int \dfrac{dx}{\sqrt{(25 - 9x^2)}}$;   (o) $\int \dfrac{dx}{1 + 3x^2}$.

2. Use integration by parts to find the following indefinite integrals:

(a) $\int x e^x dx$;   (b) $\int x \cos x dx$;   (c) $\int x \ln x dx$;

(d) $\int x^2 \sin x dx$;   (e) $\int e^x \sin x dx$;   (f) $\int x(4x - 3)^{\frac{3}{2}} dx$.

3. Use partial fractions to find the following indefinite integrals.

(a) $\int \dfrac{dx}{(x-1)(x-2)}$;   (b) $\int \dfrac{x^2+2x+3}{x-3} dx$;   (c) $\int \dfrac{x^2-x+1}{x^2-5x+6} dx$;

(d) $\int \dfrac{dx}{(x-1)(x-2)^2}$;   (e) $\int \dfrac{2x-3}{(x^2+1)(x-1)} dx$; (f) $\int \dfrac{x^2-2x+5}{(x^2-3)(x+2)^3} dx$.

4. Evaluate the following integrals.

(a) $\int_0^2 x dx$;   (b) $\int_0^1 (1-x^2) dx$;   (c) $\int_0^{\frac{\pi}{4}} \sin x dx$;

(d) $\int_0^1 e^{-x} dx$;   (e) $\int_{\frac{\pi}{4}}^{\frac{\pi}{2}} x \sin x dx$.

5. The rate at which a population is growing is given by

$$\frac{dP}{dt} = 13.75e^{0.025t}$$

where t is the time in hours. A quadratic approximation to this is

$$\frac{dP}{dt} \approx 13.75(1 + 0.025t + 0.000625t^2).$$

Find by how much the populations based upon each·of these two growth rates differ when t is 10, 100, and 1000 hours.

6. The rate at which solar energy is falling upon a horizontal surface is 720 $\sin((\pi/12)t)$ W m^{-2} where t is in hours and covers a 12-hour period from sunrise to sunset. How much energy does each square metre of surface receive over this twelve-hour period?

5.6. Numerical integration

Where an integration is required and it has not proved possible to do this analytically it becomes necessary to employ numerical techniques. These numerical methods can only be applied to finding specific areas because they deal with actual numbers rather than variables. This means that they can be used to evaluate definite integrals but cannot help at all in finding an indefinite integral. In any practical situation it is usually specific evaluations that are required and so these methods are in wide use.

In §5.2 the area under a curve was approximated by dividing it up into a number of vertical strips. The crudest approximation is to take a rectangular strip but with all such methods the accuracy improves as more and more strips are taken and their widths decrease correspondingly. It was shown that as the number of strips tended to infinity and the strip width tended to zero the accuracy became absolute. In a practical situation the calculation of the areas of a near infinite number of rectangular strips is not possible. For this reason a better approximation in the shape of the strip is employed. An equivalent accuracy can then be attained with fewer strips.

A shape improvement over the rectangular strip (Figure 5.21(a)) is to use a trapezoidal strip (Figure 5.21(b)) in which the curve is approximated to by a number of chords. This particular approximation was used in §5.4 to estimate the area under a section of $y = 1/x$. A fixed strip width, or step length, h, is used, determined by the

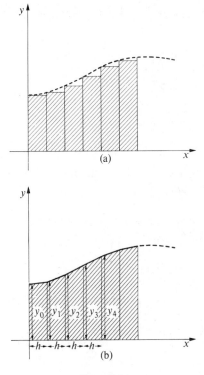

FIG. 5.21

number of strips, n, chosen to cover the required area. The height of each side of every trapezium needs to be known in order to calculate its area and so a regular set of experimental points is required or, if the function associated with the graph is known, a set of points must be calculated.

It will be assumed that the area required lies between $x = a$ and $x = b$ where $b > a$. The n strips will therefore have width $h = (b - a)/n$. Associated with these strips will be $(n + 1)$ values of x which define the sides. If these are designated by $x_0, x_1, x_2, \ldots, x_n$ and are all equally spaced between $x = a$ and $x = b$, a general point x_i will be given by

$$x_i = a + ih, i = 0, 1, \ldots, n$$

so that $x_0 = a$ and $x_n = a + nh = b$. The corresponding y values will be $y_0, y_1, y_2, \ldots, y_n$ which give the heights of the sides.

The first trapezium has area (Figure 5.21(b))

$$A_1 = \frac{y_0 + y_1}{2} \times h = \frac{1}{2}h(y_0 + y_1),$$

the second

$$A_2 = \frac{1}{2}h(y_1 + y_2),$$

and the ith

$$A_i = \frac{1}{2}h(y_{i-1} + y_i).$$

The sum of all such areas will give the required approximation to the area between $x = a$ and $x - b$.

$$A = \frac{1}{2}h(y_0 + y_1) + \frac{1}{2}h(y_1 + y_2) + \frac{1}{2}h(y_2 + y_3) + \ldots + \frac{1}{2}h(y_{n-1} + y_n)$$

$$= \frac{1}{2}h\{y_0 + y_n + 2(y_1 + y_2 + \ldots + y_{n-1})\}.$$

This is known as the *trapezoidal rule* for numerical integration.

Example. Find an approximation to

$$\int_1^3 \frac{dx}{x}$$

using the trapezoidal rule with 10 strips.

With 10 strips from 1 to 3 the strip width is 0.2. The curve is $y = 1/x$ and so the following points are calculated.

Suffix	x	y
0	1.0	1.00000
1	1.2	0.83333
2	1.4	0.71429
3	1.6	0.62500
4	1.8	0.55556
5	2.0	0.50000
6	2.2	0.45455
7	2.4	0.41667
8	2.6	0.38462
9	2.8	0.35714
10	3.0	0.33333

Sum (suffixes 1–9) is 4.84116

Hence, the approximation is

$$\frac{1}{2} \times 0.2 \times (1.00000 + 0.33333 + 2 \times 4.84116)$$

$$= 1.101565.$$

In most cases the graph passing through a set of points will be a curve and so a better approximation to the area under this curve would be to use strips which have an appropriately curved top. The simplest curve is a parabola and if sections of a parabola can be matched to the set of points it can be expected that the calculated area will be more accurate than that obtained from straight-line approximations. The trapezoidal rule uses a number of linear approximations to the curve and the method being considered now employs quadratic approximations.

A parabola is defined by three points (§2.3) and so two strips are required for each section of parabola. This means that the area being calculated must be divided into an *even* number of strips. The first two strips are shown in Figure 5.22 and are defined by the points (x_0, y_0), (x_1, y_1) and (x_2, y_2). We require a parabola,

$$y = a_0 + a_1 x + a_2 x^2,$$

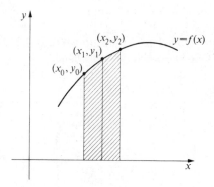

$$\text{Fig. 5.22}$$

which passes through each of these points to represent the curve passing through these points. The integral of this parabola from $x = x_0$ to $x = x_2$ will be the approximation we require to the area under the curve in the first two strips. Such a parabola has been found in §3.7 where a quadratic interpolation formula was derived. The parabola is

$$y = y_0 + \frac{x - x_0}{h} \Delta y_0 + \frac{(x - x_1)(x - x_0)}{2h^2} \Delta^2 y_0$$

where $\Delta y_0 = y_1 - y_0$ and $\Delta^2 y_0 = \Delta y_1 - \Delta y_0 = y_2 - 2y_1 + y_0$. We therefore require to evaluate

$$\int_{x_0}^{x_2} \left\{ y_0 + \frac{\Delta y_0}{h}(x - x_0) + \frac{\Delta^2 y_0}{2h^2}(x - x_1)(x - x_0) \right\} dx.$$

This can be simplified to some extent by letting $x - x_0 = u$. The new limits of integration are therefore 0 to $x_2 - x_0$. However, $x_2 - x_0 = 2h$ and so the integral becomes

$$\int_0^{2h} \left\{ y_0 + \frac{\Delta y_0}{h} u + \frac{\Delta^2 y_0}{2h^2}(u - h)u \right\} du$$

$$= \left\{ y_0 u + \frac{\Delta y_0}{h} \frac{1}{2} u^2 + \frac{\Delta^2 y_0}{2h^2} \left(\frac{1}{3} u^3 - h \frac{1}{2} u^2 \right) \right\} \Bigg|_0^{2h}$$

$$= 2h y_0 + 2h \Delta y_0 + \frac{1}{3} h \Delta^2 y_0$$

$$= 2hy_0 + 2h(y_1 - y_0) + \frac{1}{3}h(y_2 - 2y_1 + y_0)$$

$$= \frac{1}{3}h(y_0 + 4y_1 + y_2).$$

This is the area of the first pair of strips and replacing y_0, y_1, and y_2 by y_2, y_3, and y_4 gives the area of the second pair. The total area is therefore given by

$$A = \frac{1}{3}h(y_0 + 4y_1 + y_2) + \frac{1}{3}h(y_2 + 4y_3 + y_4) + \frac{1}{3}h(y_4 + 4y_5 + y_6) + \ldots$$

$$\ldots + \frac{1}{3}h(y_{n-2} + 4y_{n-1} + y_n)$$

$$= \frac{1}{3}h(y_0 + 4y_1 + 2y_2 + 4y_3 + 2y_4 + 4y_5 + \ldots + 2y_{n-2} + 4y_{n-1} + y_n)$$

$$= \frac{1}{3}h(y_0 + y_n + 2(y_2 + y_4 + \ldots + y_{n-2}) + 4(y_1 + y_3 + \ldots + y_{n-1})).$$

This is *Simpson's rule* for numerical integration. It requires an even number of strips and is generally more accurate than the trapezoidal rule.

Example. Find an approximation to

$$\int_1^3 \frac{dx}{x}$$

using Simpson's rule with 10 strips.

The table of values has been calculated in the previous example. With 10 strips Simpson's rule is written

$$A = \frac{1}{3}h \ \{y_0 + y_{10} + 2(y_2 + y_4 + y_6 + y_8) + 4(y_1 + y_3 + y_5 + y_7 + y_9)\}$$

and the table of values gives

$$A = \frac{1}{3} \times 0.2 \times \{1.33333 + 2 \times 2.10902 + 4 \times 2.73214\}$$

$$= 1.098662.$$

The integral can be evaluated exactly to give

$$\int_1^3 \frac{\mathrm{d}x}{x} = \ln 3 = 1.098612$$

and so it is possible to compare the effectiveness of these two methods on this calculation.

		Error	
Exact value	ln 3 = 1.098612		
Trapezoidal rule	1.101565	0.002953	0.27 per cent
Simpson's rule	1.098662	0.000050	0.0046 per cent

Both methods used 10 strips each of width 0.2. The greater accuracy of Simpson's rule means that for a specified degree of accuracy in the calculation fewer strips are required with Simpson's rule than with the trapezoidal rule. In fact, using 4 strips, each of width 0.5, Simpson's rule gives 1.10000 as the area which is in error by 0.001388 or 0.13 per cent. This error is still half the size of that due to the trapezoidal rule using 10 strips.

The methods are well adapted for computers as the following example using Simpson's rule demonstrates. The integral to be evaluated is

$$\int_0^1 \frac{\mathrm{d}x}{1 + x^2}.$$

The program in BASIC requests the lower and upper limits and the number of strips required.

```
10 PRINT "SIMPSON'S RULE FOR 1/(1 + X ↑ 2)": PRINT
20 DEF FNA(X) = 1/(1 + X ↑ 2): REM DEFINITION OF FUNCTION
30 INPUT "INPUT LOWER AND UPPER LIMITS L1, L2", L1, L2: PRINT
40 INPUT "INPUT NUMBER OF STRIPS N", N: PRINT
50 H = (L2 − L1)/N: X = 0; Y = 0: Z = 0
60 X = FNA(L1) + FNA(L2)
70 FOR I = 2 to N − 2 STEP 2
80 Y = Y + FNA(L1 + I * H)
90 NEXT I
100 FOR I = 1 to N − 1 STEP 2
110 Z = Z + FNA(L1 + I * H)
120 NEXT I
130 A = (H/3) * (X + 2 * Y + 4 * Z)
```

140 PRINT "VALUE OF INTEGRAL IS"; A; PRINT
150 END

On running the program the following print out is produced.

> SIMPSON'S RULE FOR $1/(1 + X \uparrow 2)$
> INPUT LOWER AND UPPER LIMITS L1, L2? 0, 1
> INPUT NUMBER OF STRIPS N? 10
> VALUE OF INTEGRAL IS .7853981534843
> END PROGRAM

The integral can be evaluated exactly to give $\frac{1}{4}\pi$ (= 0.7853981633975
. . .) and so the error is -1.3×10^{-6} per cent. To evaluate a different
integral, statements 10 and 20 require an appropriate change.

Exercises

1. Use the trapezoidal rule and Simpson's rule, each with 10 strips,
to evaluate:

(a) $\int_1^4 \frac{dx}{x}$;   (b) $\int_0^\pi \sin x \, dx$;   (c) $\int_2^3 x \, e^{-x} dx$;

(d) $\int_2^4 \frac{x}{x^2 - 1} \, dx$;   (e) $\int_{\frac{\pi}{4}}^{\frac{\pi}{3}} x^2 \cos x \, dx$.

Evaluate each integral exactly and compare the numerical results with
the exact result in each case.

2. Sketch the *standard normal curve*

$$y = \frac{1}{\sqrt{(2\pi)}} e^{-\frac{x^2}{2}}$$

If the area beneath this curve from $x = -\infty$ to $x = -5$ can be neglected
evaluate

$$\frac{1}{\sqrt{(2\pi)}} \int_{-\infty}^x e^{-\frac{t^2}{2}} \, dt$$

for $x = 0, 1, 2, 3, 4$, and 5 using a step length of 0.2.

3. Recordings of the rate at which solar energy in W m^{-2} fell upon a
horizontal surface over a ten-hour period on a hazy day were found
to be

Time (h)	0	1	2	3	4	5	6	7	8	9	10
Energy rate (W m^{-2})	3	152	289	395	432	421	384	372	277	148	5

Find the total amount of energy which fell upon 1 m^2 of horizontal surface during this period.

6 Differential equations

6.1. Occurrence and Modelling

BEFORE becoming involved in the solution of differential equations it is perhaps as well to consider some of the situations in which such equations occur. For this purpose a differential equation may be regarded as an equation containing one or more derivatives. More detailed definitions and a classification are given in §6.2.

Any derivative implies a rate of change of a dependent variable with respect to an independent variable. Hence, any situation in which a rate of change is known to depend upon quantifiable factors is a candidate for expression in terms of differential equations. There are numerous examples of this in biology, chemistry, and physics, all of interest to the biological scientist. As illustrations consider the following examples.

Simple Population Growth. In a favourable environment a biological population grows at a rate proportional to its size at any time. If the population is sufficiently large this growth may be regarded as a continuous rather than a discrete process.

In more analytic terms we may say

Rate of increase of population = constant of proportionality
                                    × size of population

or, mathematically

$$\frac{\mathrm{d}P}{\mathrm{d}t} = kP$$

where P is the size of the population at any time t and k a constant of proportionality. Note that since the population is increasing $\mathrm{d}P/\mathrm{d}t$ must be positive so that k must also be positive.

The growth of the population is described by the above differential equation. The solution should give the population P as a function of time t. However, without solving the equation, it is possible to sketch the variation of P with t. If initially P is zero $\mathrm{d}P/\mathrm{d}t$ remains zero for all time and there can be no growth. We must therefore assume that the initial value of P, P_0 say, is greater than zero. This enables us to plot the initial point on the graph. In addition, from the differential equation, we know that the slope of the graph through this point is kP_0. If at some time T later the population has risen to $2P_0$, as certainly it must if it is continually increasing at a faster and faster rate, then the slope at this point is $2kP_0$. With just this information a rough sketch of P against t can be made (Figure 6.1). Obviously the precise shape of the curve depends upon k but the general shape will always be as above.

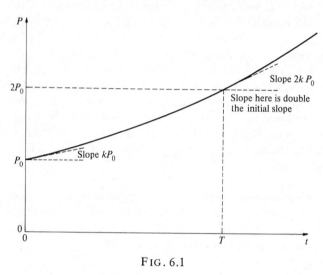

F IG . 6.1

Note that in order to draw this curve the initial value of P had to be known, or assumed. Such extra information is always required to solve differential equations describing specific problems.

Radioactive decay. A radioactive isotope decays at a rate proportional to its mass at any time.

This is very similar to the previous problem but in this case there is a continuous decrease. A similar argument leads to

$$\frac{dm}{dt} = -km$$

where m is the mass at time t and k is positive. The minus sign is required to ensure that m is decreasing with time i.e. $dm/dt < 0$.

Again a rough sketch of m this time against t can be made if it is assumed that $m = m_0$ when $t = 0$ and that at some subsequent time T the mass has decreased to half its initial value. This time T is known as the *half-life* of the isotope concerned. In every period T it has decayed to half the mass it had at the beginning of the period (Figure 6.2).

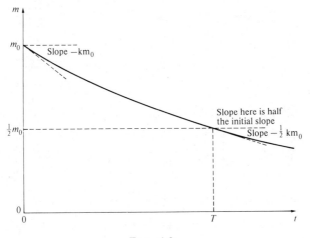

FIG. 6.2

Newton's law of cooling. Under conditions of forced convection the rate at which a body cools is proportional to the difference in temperature between the body and the cooling fluid.

Cooling implies heat loss but for a given body this is proportional to temperature loss. Newton's law of cooling may therefore be expressed mathematically by

$$\frac{\mathrm{d}T}{\mathrm{d}t} = -k(T - T_c)$$

where T is the temperature at time t, and T_c the temperature (assumed to be constant) of the cooling fluid and k is a positive constant of proportionality. Note that if $T < T_c$ then $\mathrm{d}T/\mathrm{d}t > 0$ and temperature increases with time. In either case the temperature approaches T_c. The nearer T gets to T_c the smaller the slope (Figure 6.3).

The law of mass action. At constant temperature the rate of a chemical reaction is proportional to the product of the concentrations of the reacting substances.

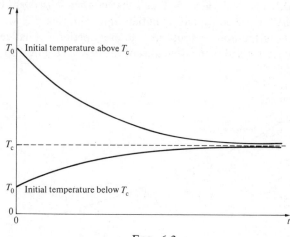

FIG. 6.3

To be more specific consider the unimolecular reaction

$$X + Y \rightarrow Z$$

in which 1 mole of X reacts with 1 mole of Y to give 1 mole of Z. In addition, let the initial concentration of X be x_0 moles per litre and of Y be y_0 moles per litre and assume there is no Z. The rate of production of Z is a measure of the reaction rate and if at some subsequent time z moles per litre of Z have been produced this has used z moles per litre of X and z moles per litre of Y. At this stage the concentrations of X and Y are $(x_0 - z)$ and $(y_0 - z)$ respectively and the reaction rate is proportional to their product. However, dz/dt is proportional to the reaction rate and so

$$\frac{dz}{dt} = k(x_0 - z)(y_0 - z).$$

In this case it is much more difficult to sketch a graph of z against t. However, it will be noted that z increases with time and levels off as z approaches the lower of x_0 and y_0.

Bounded population growth. A population cannot keep on growing at an ever-increasing rate. It will eventually be limited by available food supplies, space, or some other feature. The simple population growth does not take account of this. The simple model can however represent the growth provided the population is much lower than the maximum supportable by a given environment. In such conditions the environment can be regarded as favourable. As the population grows the environment becomes less favourable and it is to be expected that the population will start to level out as it approaches the maximum that can be supported. Such levelling out can be represented by

$$\frac{\mathrm{d}P}{\mathrm{d}t} = k(P_M - P)$$

where P_M is the maximum supportable population. As P approaches P_M, $\mathrm{d}P/\mathrm{d}t$ tends to zero and so the graph levels out (Figure 6.4).

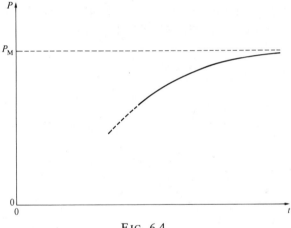

FIG. 6.4

There are now two differential equations,

$$\frac{\mathrm{d}P}{\mathrm{d}t} = kP, \qquad \frac{\mathrm{d}P}{\mathrm{d}t} = k(P_M - P),$$

the first representing growth in the early stages, provided the initial population P_0 is much less than P_M, and the second growth in the final stages. These two can be combined in the single equation, known as the *logistic differential equation*,

$$\frac{\mathrm{d}P}{\mathrm{d}t} = kP(P_M - P).$$

This reproduces the features of each of the two previous equations in their particular areas of applicability. Initially P is small compared with P_M and so the equation approximates to

$$\frac{dP}{dt} = kPP_M$$

which is simple growth. As P approaches P_M the equation is approximately

$$\frac{dP}{dt} = kP_M(P_M - P)$$

which levels out at P_M. Joining these two types of growth together gives the curve in Figure 6.5. This represents the observed growth quite well in a bacterial culture for example. The solution of this differential equation leads to the *logistic growth curve*.

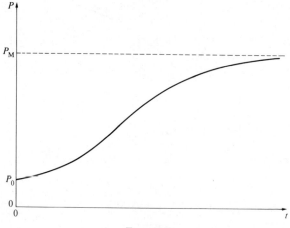

FIG. 6.5

Periodic phenomena. Heartbeats, respiration, brain waves, and so on are all periodic phenomena as are seasonal variations in populations, the length of the day, and migrations. All such patterns can be built up by a process known as Fourier synthesis from a mixture of sine and cosine waves. The study of an electroencephalogram, for instance, can therefore be reduced to a study of sine and cosine waves involving rates of change with respect to time.

In order to have some idea of when to expect periodic phenomena to occur it would be useful to know what type of differential equation has a sine or cosine wave as its solution. To do this assume that we are looking for a differential equation with a periodic solution

$$x = A \sin(\omega t)$$

where A is a constant, known as the amplitude, and ω is another constant, known as the angular frequency. Then

$$\frac{dx}{dt} = A\omega \cos(\omega t)$$

$$\frac{d^2x}{dt^2} = -A\omega^2 \sin(\omega t) = -\omega^2 x.$$

Hence, the differential equation with the above solution is

$$\frac{d^2x}{dt^2} + \omega^2 x = 0.$$

Substitution shows that

$$x = B \cos(\omega t)$$

and

$$x = A \sin(\omega t) + B \cos(\omega t)$$

are also solutions of this equation. Note that the last solution is the sum of the other two. The precise form of the solution is determined by extra information such as the initial values of x and dx/dt.

Exercises

1. The growth of a particular population may be represented by the differential equation

$$\frac{dP}{dt} = 0.07P.$$

If the population is initially one million and doubles in about 10 hours, sketch the variation of the population P with time t.

2. A population which is initially of size P_0 doubles every two hours. Sketch the variation of population P with time t. Does the equation

$$P = P_0(2)^{\frac{t}{2}}$$

represent the variation of P with t? Does this equation satisfy a differential equation of the form

$$\frac{dP}{dt} = kP?$$

(Hint: Note that 2 may be written $e^{\ln 2}$.)

3. The radioactive isotope Carbon 14 (C^{14}) has a half-life of 5570 years. If initially there are 2 mg of this isotope write down an equation for the amount remaining at any subsequent time t. Does this equation satisfy a differential equation of the form

$$\frac{dm}{dt} = -km ?$$

4. Sometimes it is found that a population must be greater than a critical size P_c before it grows. The behaviour of such a population may be represented by a differential equation of the form

$$\frac{dP}{dt} = k(P - P_c)(P_M - P).$$

If the initial size P_0 of the population is assumed to be much less than P_M sketch graphs of the variation of P with t for the three cases $P_0 > P_c$, $P_0 = P_c$, and $P_0 < P_c$.

6.2. Definitions

The previous section gave an indication of the variety of differential equations which can occur. It will probably come as no surprise to learn that differential equations which look similar are solved in similar ways. In fact the equations are classified into various types which have similar mathematical characteristics.

A starting point can be made from the statement that

any equation which contains derivatives of an unknown function
is a differential equation.

This can be taken as a definition. All of the following examples would be included.

1. $\dfrac{dy}{dx} = 4x^2$;

2. $\dfrac{dy}{dx} + 2xy = e^x$;

3. $\dfrac{dy}{dx} + xy^2 = \cos x$;

4. $\left(\dfrac{dy}{dx}\right)^3 + \cos xy^4 = e^{2x} \sin 3x$;

5. $\dfrac{d^2y}{dx^2} + 4y = 0$;

6. $\dfrac{d^2y}{dx^2} + 3\dfrac{dy}{dx} + 5y = \sin 4x$;

7. $\dfrac{d^2y}{dx^2} + 4\left(\dfrac{dy}{dx}\right)^2 + y = 3x^2$;

8. $\dfrac{d^3y}{dx^3} + 2x^2\dfrac{dy}{dx} + \cos y = \sin x$;

9. $\left(\dfrac{d^2x}{dt^2}\right)^4 + 4t^2x = 0$;

10. $\dfrac{\partial C}{\partial t} = D\dfrac{\partial^2 C}{\partial x^2}$.

The initial classification is into ordinary and partial.

An *ordinary differential equation* (O.D.E.) is an equation containing one or more derivatives of a dependent variable with respect to an independent variable. It may also contain functions of both of these variables. Of the above examples 1 to 9 are ordinary differential equations. It is this type of equation that is considered in this chapter. Example 10 is a *partial differential equation* (P.D.E.). Such equations are beyond the scope of this book. However, it is perhaps worth noting that some partial differential equations can be broken down into two or more ordinary differential equations.

Differential equations are further classified by order and degree. The *order* of a differential equation is the order of the highest derivative occurring in it. The *degree* is the power to which this highest derivative is raised. This power is assumed to be a positive integer. Of the above examples 1, 2, 3, and 4 are first order, 5, 6, 7, 9, and 10 are second order, and 8 is third order. All of the examples are of the first degree except for 4 which is third and 9 which is fourth.

A further, but very important, subclassification is by linearity. A differential equation is *linear* if, when the *dependent* variable and its derivatives occur, they do so only in the first degree and then not in products with each other. Note that nothing is said about the occurrence of the *independent* variable. If an equation is not linear it is said to be *non-linear*. In general non-linear equations are much more difficult to solve than linear equations. Of the above examples 1, 2, 5, 6, and 10 are linear.

Exercises

1. State the order and degree of the following differential equations and indicate which are linear.

(a) $\dfrac{dy}{dx} = 0$;

(b) $\dfrac{d^2y}{dx^2} + 2\dfrac{dy}{dx} - 3y = 0$;

(c) $\dfrac{d^2y}{dx^2} + 4y = x^2$;

(d) $\dfrac{dy}{dx} + x^2 y = e^x$;

(e) $\dfrac{d^2y}{dx^2} + \left(\dfrac{dy}{dx}\right)^2 + 3y = \cos x$;

(f) $\dfrac{d^2y}{dx^2} = y^2$;

(g) $\dfrac{d^3x}{dt^3} = te^t$;

(h) $\dfrac{dy}{dx} = 2x^2 - 4y$;

(i) $\dfrac{dy}{dx} = 2x(x - 2y)$;

(j) $\left(\dfrac{dy}{dx}\right)^2 + 2x\dfrac{dy}{dx} + 3x^2y^2 = 0$;

(k) $\dfrac{dy}{dx} + \cos y = x$;

(l) $\dfrac{d^2y}{dt^2} + 2t^3\dfrac{dy}{dt} - 3(t-1)y = \sin t$.

6.3. First-order separable

Ideally the solution of a differential equation should present the dependent variable as a function of the independent variable. Unfortunately this cannot always be accomplished and so in general any relationship involving the dependent and independent variables from which derivatives have been eliminated and which satisfies the differential equation can be regarded as a solution.

The easiest equations to solve are those first-order equations in which the dependent and independent variables can be put on opposite sides of the equals sign. Such equations are said to be *separable* and are solved by direct integration. The simplest form is

$$\frac{dy}{dx} = f(x).$$

Separation then gives

$$\int dy = \int f(x)dx + C,$$

or

$$y = F(x) + C,$$

where $F(x)$ is the indefinite integral of $f(x)$. C is an arbitrary constant of integration and so there are infinitely many solutions of the differential equation, each differing from any other by a constant. The *solution curves*, that is graphs of the solutions, will all look alike but be displaced vertically from one another, the displacement depending upon the value of the arbitrary constant chosen.

Example. Solve the differential equation

$$\frac{dy}{dx} = 2x - 1$$

and sketch some solution curves.

Separation gives

$$\int dy = \int (2x - 1)dx + C$$

or

$$y = x^2 - x + C.$$

Various solution curves can be drawn by choosing different values for
the arbitrary constant C (Figure 6.6).

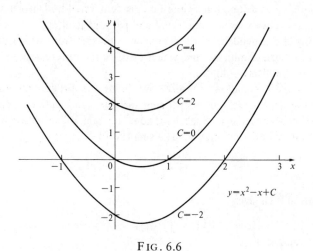

FIG. 6.6

If a particular solution curve is required it must be specified by
giving some further information e.g. the solution of $y' = 2x - 1$ which
passes through the origin. Such information determines the constant C.

The more general separable equation has the form

$$\frac{\mathrm{d}y}{\mathrm{d}x} = \frac{f(x)}{g(y)}$$

so that

$$\int g(y)\mathrm{d}y = \int f(x)\mathrm{d}x + C$$

$$G(y) = F(x) + C$$

where $F(x)$ and $G(y)$ are the indefinite integrals of $f(x)$ and $g(y)$,
respectively. Again there are an infinite number of solutions, each one
determined by the value of C chosen.

Example. Solve the simple population growth equation

$$\frac{\mathrm{d}P}{\mathrm{d}t} = kP.$$

If the population is initially 200 000 and 4 hours later has grown to 300 000, find the population P as a function of time t.

The equation is separable, though not quite of the previous type. It may be integrated.

$$\int \frac{dP}{P} = k \int dt + C$$

or

$$\ln P = kt + C$$

so that

$$P = e^{kt+c} = e^c e^{kt}.$$

C is the constant of integration and so again there are an infinite number of solutions. Since C is a constant so is e^c. Note that when $t = 0$, $P = e^c$ and so e^c represents the initial size of the population. Denote this by P_0. Therefore,

$$P = P_0 e^{kt}.$$

Specifying a particular initial population defines a particular solution curve. It is that curve which passes through $(0, P_0)$ on a graph of P against t.

Initially the population is 200 000 and so $P_0 = 200\,000$. The additional information that when $t = 4$, $P = 300\,000$ gives a means of evaluating k since

$$300\,000 = 200\,000e^{4k}.$$

$$e^{4k} = 1.5,$$

$$4k \log e = \log 1.5,$$

$$k = \tfrac{1}{4} \frac{\log 1.5}{\log e} = \frac{0.1761}{4 \times 0.4343} = 0.1014,$$

and so

$$P = 200\,000e^{0.1014t}.$$

Example. Solve

$$\frac{dy}{dx} = 2x \sec y, \quad y = 0 \text{ when } x = 0.$$

Separation gives

$$\int \frac{dy}{\sec y} = \int 2x dx + C.$$

Now

$$\int \frac{dy}{\sec y} = \int \cos y \, dy = \sin y$$

and so

$$\sin y = x^2 + C.$$

Since $y = 0$ when $x = 0$ the constant C is zero and a particular solution curve is specified.

Example. Diffusion of a compound through a membrane is described by the partial differential equation (cf. §6.2)

$$\frac{\partial C}{\partial t} = D \frac{\partial^2 C}{\partial x^2},$$

where C is the concentration of the compound, t is time, x is distance into the membrane, and D is a constant known as the diffusion coefficient. When a steady state is reached there is, by definition, no change with time and so $\partial C/\partial t = 0$ and the partial differential equation reduces to the ordinary differential equation

$$\frac{d^2 C}{dx^2} = 0.$$

If the membrane is of thickness a and the concentration on one side is C_1 and on the other it is C_2, find the concentration C at any point inside the membrane after a steady state has been reached.

The second order equation can be written

$$\frac{d}{dx}\left(\frac{dC}{dx}\right) = 0$$

and by putting $dC/dx = R$ say, a new variable, the second order equation becomes two first order equations,

$$\frac{dR}{dx} = 0 \text{ and } \frac{dC}{dx} = R.$$

The first equation is separable and will yield R which can then be substituted into the second equation. The solution of the first equation is

$$R = A, \text{a constant}$$

and so

$$\frac{dC}{dx} = A \quad \text{or} \quad C = Ax + B$$

after integration. The constants A and B can be found from the conditions imposed on each side of the membrane. If the distance x through the membrane is measured from the side at which the concentration is C_1 then $C = C_1$ at $x = 0$ and $C = C_2$ at $x = a$. Substitution of these conditions into the solution gives

$$C_1 = B$$

and

$$C_2 = Aa + B.$$

Hence

$$B = C_1 \quad \text{and} \quad A = \frac{1}{a}(C_2 - C_1)$$

and so

$$C = (C_2 - C_1)\frac{x}{a} + C_1.$$

The concentration therefore changes linearly from C_1 to C_2 as the membrane is crossed.

Exercises

1. Solve

(a) $\dfrac{dy}{dx} = x$;

(b) $\dfrac{dy}{dx} = y$, $y = 1$ when $x = 0$;

(c) $\dfrac{dy}{dx} = \dfrac{x^3}{(y+1)^2}$, $y = 0$ when $x = 0$;

(d) $\dfrac{dx}{dt} = e^{t-x}$;

(e) $\dfrac{dy}{dx} = \dfrac{y(x^2-1)}{x^2(y^2-y)}$, $y = 4$ when $x = 2$.

2. A population is growing at a rate proportional to its size at any time. When first observed the population was 2 million and two days later it had risen to 3 million. Find an expression for the size of the population as a function of time and hence determine its size after five further days.

3. The ratio of the mass of the radioactive isotope C^{14} to that of the stable isotope C^{12} in a living organism is constant. When the organism dies the isotope C^{14} starts to decay at a rate proportional to the mass present while the mass of the isotope C^{12} remains constant. The proportion of C^{14} to C^{12} present compared with that in a living organism can therefore be used to find out how long ago an organism died. If after t years the mass of C^{14} present is x per cent of that in a living organism express t as a function of x. The half-life of C^{14} is 5570 years.

4. A gas at $50\,^{\circ}$C is blown past a body which was initially at $100\,^{\circ}$C. After 30 minutes the temperature of the body has dropped to $70\,^{\circ}$. Find how long it takes to drop to $60\,^{\circ}$C. If the initial temperature of the body had been $20\,^{\circ}$C, find what its temperature would have been after 1 hour in the gas flow.

5. The rate of change of a population of size P with an upper bound P_M may be represented by the logistic differential equation

$$\frac{dP}{dt} = kP(P_M - P).$$

A population has an upper bound of 90 million and in the last 10 years has risen from 10 million to 30 million. Find the size of the population as a function of time from the present and hence predict its size in 10 years time.

6. A compound is diffusing through the wall of a cylindrical blood vessel. Once a steady state has been reached the differential equation governing the process is

$$\frac{d^2C}{dr^2} + \frac{1}{r}\frac{dC}{dr} = 0,$$

where C is the concentration of the compound and r is the radial distance outwards from the axis of the blood vessel. If the concentration at the internal wall, radius r_1, of the blood vessel is C_1 and at the external wall, radius r_2, it is C_2, find an expression for the concentration C at any radial distance r within the wall ($r_1 \leqslant r \leqslant r_2$).

6.4. Types convertible to separable form

Some first-order equations which are not separable can be converted to separable form by choice of a suitable change of variable. As with integration this choice is made easier as experience is accumulated. However, there are one or two types of first-order equation where the choice is fairly obvious.

Consider the equation

$$\frac{dy}{dx} = f(ax + by).$$

The right-hand side is simplified if a new variable is introduced such that

$$v = ax + by.$$

Differentiating this with respect to x gives

$$\frac{dv}{dx} = a + b\frac{dy}{dx}$$

but with the new variable

$$\frac{dy}{dx} = f(v)$$

and so the new differential equation is

$$\frac{dv}{dx} = a + bf(v)$$

which is separable.

Example. Solve $dy/dx = x + y$, $y = 1$ when $x = 0$.

To simplify the right-hand side let $v = x + y$ then

$$\frac{dv}{dx} = 1 + \frac{dy}{dx} = 1 + v.$$

Therefore,

$$\int\frac{dv}{1 + v} = \int dx + C,$$

$$\ln|1 + v| = x + C,$$

$$1 + v = Ae^x.$$

Hence,

$$y = Ae^x - x - 1.$$

Since $y = 1$ when $x = 0$ the value of A is 2.

$$y = 2e^x - x - 1.$$

If a first-order equation can be written in the form

$$\frac{dy}{dx} = f\left(\frac{y}{x}\right)$$

it is said to be *homogeneous*. Such an equation can be reduced to separable form by the substitution

$$v = \frac{y}{x}.$$

Differentiating $y = vx$ shows that $dy/dx = v + x(dv/dx)$ and the equation becomes

$$v + x\frac{dv}{dx} = f(v)$$

or

$$x\frac{dv}{dx} = f(v) - v$$

which is separable.

Example. Solve

$$\frac{dy}{dx} = \frac{2xy}{x^2 - y^2}.$$

This can be written

$$\frac{dy}{dx} = \frac{2y/x}{1 - (y/x)^2}$$

which is homogeneous. The substitution, $y = vx$, then gives

$$v + x\frac{dv}{dx} = \frac{2v}{1 - v^2}.$$

Hence,

$$x\frac{dv}{dx} = \frac{2v}{1 - v^2} - v = \frac{v(1 + v^2)}{1 - v^2},$$

$$\int \frac{1 - v^2}{v(1 + v^2)}\, dv = \int \frac{dx}{x} + C.$$

Partial fractions gives

$$\int \left(\frac{1}{v} - \frac{2v}{1 + v^2}\right) dv = \ln|x| + C,$$

and so

$$\ln|v| - \ln|1 + v^2| = \ln|x| + C$$

or

$$\frac{v}{v^2 + 1} = Ax.$$

Returning to the original variables shows that

$$\frac{xy}{x^2 + y^2} = Ax \quad \text{or} \quad x^2 + y^2 = \frac{1}{A}y.$$

Exercises

1. By a suitable change of variable convert the following differential equations to separable form and hence solve them.

(a) $\dfrac{dy}{dx} = x - y$;   (b) $\dfrac{dy}{dx} = 4y - 3x$;   (c) $\dfrac{dy}{dx} = (x + y)^2$;

(d) $\dfrac{dy}{dx} = (x - y)^2$, $y = 0$ when $x = 0$;

(e) $\dfrac{dy}{dx} = \tan^2(x + y)$, $y = 0$ when $x = 0$.

2. Show that the change of variable $v = x^2 + y^2$ reduces the differential equation

$$x + y\frac{dy}{dx} = x^2 + y^2$$

to separable form and hence solve it.

3. Show that the equation

$$\frac{dy}{dx} = f(ax + by + c)$$

may be reduced to separable form by a suitable substitution.

4. Solve the following homogeneous equations.

(a) $\dfrac{dy}{dx} = \dfrac{x + y}{x}$;   (b) $\dfrac{dy}{dx} = \dfrac{x + y}{x - y}$;   (c) $\dfrac{dy}{dx} = \dfrac{x^2 + y^2}{xy}$;

(d) $\dfrac{dy}{dx} = \dfrac{x^2 - y^2}{xy}$, $y = 0$ when $x = 1$;

(e) $x\dfrac{dy}{dx} = x \tan \dfrac{y}{x} + y$, $y = \dfrac{\pi}{2}$ when $x = 1$.

6.5. First-order linear

The most general form of the *first-order linear* ordinary differential equation is

$$\frac{dy}{dx} + p(x)y = q(x)$$

where $p(x)$ and $q(x)$ are functions of the independent variable x. This type of equation can be solved by the use of an *integrating factor* which, when applied to the whole equation, enables the left-hand side to be expressed as an exact derivative and hence to be integrated directly. This eliminates the derivative and so technically the equation is solved. That such factors exist can be seen from the following example. Consider the equation

$$\frac{dy}{dx} + \frac{1}{x}y = x^2.$$

This is clearly first-order linear. If the equation is multiplied through by the factor x, the left-hand side can be expressed as an exact derivative,

$$x\frac{dy}{dx} + y = x^3,$$

$$\frac{d}{dx}(xy) = x^3.$$

Therefore,
$$xy = \int x^3 dx + C$$

$$= \frac{1}{4}x^4 + C.$$

Hence,
$$y = \frac{1}{4}x^3 + \frac{C}{x}.$$

The factor x is therefore an integrating factor which facilitates the integration of the left-hand side. A further important point to notice is that if the constant of integration had been omitted the whole term C/x, which is *not* constant, would fail to appear in the solution. *Constants of integration cannot be neglected.*

We would obviously like to know whether an integrating factor can always be found and, if so, what form it takes. In order to investigate this assume the existence of an integrating factor $I(x)$, a function of x. If this assumption leads to an inconsistency it must have been incorrect. If however no such inconsistency occurs and we are able to solve the equation, perhaps under certain conditions on $I(x)$, then the assumption is justified and the conditions necessary for the process to work may lead to an explicit form of $I(x)$. We will then have demonstrated the *existence* of the integrating factor.

Consider the general first-order linear equation

$$\frac{dy}{dx} + p(x)y = q(x)$$

and multiply through by the factor $I(x)$.

$$I(x)\frac{dy}{dx} + I(x)p(x)y = I(x)q(x).$$

If this factor is an *integrating* factor, which is the assumption, it must convert the left-hand side into an exact derivative of the product of $I(x)$ with y, i.e.

$$I(x)\frac{dy}{dx} + I(x)p(x)y = \frac{d}{dx}(I(x)y).$$

Expanding the derivative gives

$$I(x)\frac{dy}{dx} + I(x)p(x)y = I(x)\frac{dy}{dx} + y\frac{d}{dx}(I(x)).$$

Comparing terms on each side shows that

$$\frac{d}{dx}(I(x)) = I(x)p(x)$$

which is a first-order separable equation for $I(x)$.

$$\int \frac{dI(x)}{I(x)} = \int p(x)dx + C,$$

$$\ln I(x) = \int p(x)dx + C.$$

$$I(x) = e^{\int p(x)dx + C} = Ae^{\int p(x)dx}.$$

Hence, an explicit form of the integrating factor has been found. We may assume $A = 1$ without any loss of generality since we multiply the whole equation through by this constant. The solution of the original equation is straightforward.

$$I(x)\frac{dy}{dx} + I(x)p(x)y = I(x)q(x),$$

$$\frac{d}{dx}(I(x)y) = I(x)q(x),$$

$$I(x)y = \int I(x)q(x)dx + C.$$

Therefore, $$y = \frac{1}{I(x)}\int I(x)q(x)dx + \frac{C}{I(x)}$$

where

$$I(x) = e^{\int p(x)dx}.$$

Since the general form of the solution has been found the *existence* of the solution has been demonstrated. In addition the general method used to find the solution shows that it is *unique*, that is, it is the one and only possible solution. This relies on the fact that $I(x)$ is itself unique.

The concepts of, first of all, *existence* and then *uniqueness* will no doubt appear strange and rather meaningless. However, if looked at from the following point of view their importance is clear. Consider the case of a type of differential equation for which mathematicians had failed to prove the existence of a solution. All attempts at solving this type of equation might be doomed to failure because, quite simply, a solution might not exist. It is clearly important to be aware of such a possibility before too much time is wasted. Proof that a solution does exist is therefore very important. This is not always the same as finding the solution. A solution can be shown to exist without actually finding it. The other concept, that of uniqueness, can be very helpful in finding solutions. If we know that a particular type of differential equation has a unique solution then any methods we can devise, even guesswork, which produces a solution must have produced the correct solution, since, if unique, there is only one. Conversely, if there is no uniqueness, that is if a multiplicity of solutions is possible, we may not know which solution in particular is the correct one for our problem.

The solution of the equation, written in the form $y = f(x)$, consists of two parts. There is a part which depends entirely upon the form of the left-hand side of the equation and is the same whether or not there is a right-hand side at all. This is the function $1/I(x)$ and is called the *complementary function*. It is multiplied by the arbitrary constant C in the above solution. This is in fact the solution of the *reduced equation*

$$\frac{dy}{dx} + p(x)y = 0,$$

as putting $q(x) = 0$ in the original equation and its general solution shows. The other part of the solution depends upon the right-hand side of the equation, $q(x)$, and is called the *particular integral*. The general solution consists of the sum of a constant multiple of the complementary function (C.F.) and the particular integral (P.I.).

$$y = \text{constant} \times \text{C.F.} + \text{P.I.}$$

Example. Consider the original equation

$$\frac{dy}{dx} + \frac{1}{x}y = x^2.$$

The integrating factor (I.F.) is $e^{\int (1/x)dx} = e^{\ln x} = x$ as before and the solution carries through to give

$$y = \tfrac{1}{4}x^3 + \frac{C}{x}$$

where $1/x$ is the C.F. and $\tfrac{1}{4}x^3$ is the P.I.

Example.

$$\frac{dy}{dx} + 3x^2 y = 4x^3.$$

I.F. is

$$e^{\int 3x^2 dx} = e^{x^3}.$$

Then

$$e^{x^3}\frac{dy}{dx} + 3x^2 e^{x^3}y = 4x^2 e^{x^3},$$

$$\frac{d}{dx}(e^{x^3}y) = 4x^2 e^{x^3}.$$

Hence,

$$e^{x^3}y = \int 4x^2 e^{x^3}dx + C = \frac{4}{3}e^{x^3} + C.$$

Therefore,

$$y = Ce^{-x^3} + \frac{4}{3}$$

where e^{-x^3} is the C.F. and $\dfrac{4}{3}$ the P.I.

Example.

$$\frac{dy}{dx} + \frac{y}{x+2} = \sin x, \quad y = 1 \text{ when } x = 0.$$

I.F. is

$$e^{\int \frac{dx}{x+2}} = e^{\ln|x+2|} = x + 2.$$

Therefore,
$$\frac{d}{dx}((x + 2)y) = (x + 2)\sin x,$$

$$(x + 2)y = \int(x + 2)\sin x\,dx + C$$

$$= -(x + 2)\cos x + \int\cos x\,dx + C$$

$$= -(x + 2)\cos x + \sin x + C.$$

Hence
$$y = -\cos x + \frac{\sin x}{x + 2} + \frac{C}{x + 2}.$$

C.F. is $\dfrac{1}{x + 2}$ and P.I. is $-\cos x + \dfrac{\sin x}{x + 2}$.

If $y = 1$ when $x = 0$, then $C = 4$. Such extra information is always required if the arbitrary constant is to be evaluated.

Example. A population P is bounded by a seasonally varying ability of the environment to support it given by $P_M(1 - \frac{1}{4}\cos\frac{1}{2}t)$ where P_M is a constant and t is the time in months. The bounded population growth differential equation of §6.1 indicates that P is governed by an equation of the form

$$\frac{dP}{dt} = kP\{P_M(1 - \frac{1}{4}\cos\frac{1}{2}t) - P\}.$$

Discuss the solution of this equation if $k = 1/1000$, $P_M = 1600$ and the population is initially 200.

The differential equation is non-linear and non-separable. It may be rewritten as

$$\frac{dP}{dt} - \frac{8}{5}(1 - \frac{1}{4}\cos\frac{1}{2}t)P = -\frac{P^2}{1000}.$$

This is an equation of Bernoulli type (cf. exercises 6.5, question 4).

It may be converted to linear form by the substitution $z = 1/P$ so that

$$\frac{dz}{dt} = -\frac{1}{P^2}\frac{dP}{dt}.$$

Dividing the differential equation by $-P^2$ gives

$$-\frac{1}{P^2}\frac{dP}{dt} + \tfrac{8}{5}(1 - \tfrac{1}{4}\cos\tfrac{1}{2}t)\frac{1}{P} = \frac{1}{1000}$$

or

$$\frac{dz}{dt} + \tfrac{8}{5}(1 - \tfrac{1}{4}\cos\tfrac{1}{2}t)z = \frac{1}{1000}$$

which is linear. The integrating factor is

$$e^{\int \frac{8}{5}(1 - \frac{1}{4}\cos\frac{1}{2}t)dt} = e^{\frac{8}{5}(t - \frac{1}{2}\sin\frac{1}{2}t)}.$$

Multiplying through by this I.F. gives

$$\frac{d}{dt}(ze^{\frac{8}{5}(t - \frac{1}{2}\sin\frac{1}{2}t)}) = \frac{e^{\frac{8}{5}(t - \frac{1}{2}\sin\frac{1}{2}t)}}{1000}.$$

If we now integrate both sides from time $t = 0$ to some arbitrary time $t = T$ we have

$$ze^{\frac{8}{5}(t - \frac{1}{2}\sin\frac{1}{2}t)}\Big|_{t=0}^{t=T} = \frac{1}{1000}\int_0^T e^{\frac{8}{5}(t - \frac{1}{2}\sin\frac{1}{2}t)}dt.$$

No constant of integration is required since the integration is definite. Now $z = 1/P$ and initially $P = 200$. The initial value of z, z_0 say, is therefore given by $z_0 = 1/200$. If the value of z at time T is denoted by z_T, then $z_T = 1/P$. Hence

$$z_T e^{\frac{8}{5}(T - \frac{1}{2}\sin\frac{1}{2}T)} - z_0 = \frac{1}{1000}\int_0^T e^{\frac{8}{5}(t - \frac{1}{2}\sin\frac{1}{2}t)}dt.$$

The integral on the right hand side cannot be evaluated by the analytic techniques developed in Chapter 5. However, if T is specified, it can be evaluated by the numerical techniques of §5.6. We can rearrange the equation to give

$$z_T = \left\{z_0 + \frac{1}{1000}\int_0^T e^{\frac{8}{5}(t - \frac{1}{2}\sin\frac{1}{2}t)}dt\right\}e^{-\frac{8}{5}(T - \frac{1}{2}\sin\frac{1}{2}T)}.$$

If z_T and z_0 are replaced by $1/P$ and $1/200$ respectively then

$$P = \frac{e^{\frac{8}{5}(T - \frac{1}{2}\sin\frac{1}{2}T)}}{\left\{ \dfrac{1}{200} + \dfrac{1}{1000} \displaystyle\int_0^T e^{\frac{8}{5}(t - \frac{1}{2}\sin\frac{1}{2}t)}dt \right\}}$$

$$= \frac{1000\,e^{\frac{8}{5}(T - \frac{1}{2}\sin\frac{1}{2}T)}}{\left\{ 5 + \displaystyle\int_0^T e^{\frac{8}{5}(t - \frac{1}{2}\sin\frac{1}{2}t)}dt \right\}}$$

The population P may therefore be calculated at any time T even though the integral has to be evaluated numerically.

Note that at $T = 0$ the exponential in the numerator is $e^0 = 1$ and the integral in the denominator vanishes since both the upper and lower limits are zero. This gives $P = 200$, the correct initial value.

The graph of P against time is shown in Figure 6.7. The dotted curve represents the bound $P_M(1 - \frac{1}{4}\cos\frac{1}{2}t)$ with $P_M = 1600$. The graph of P

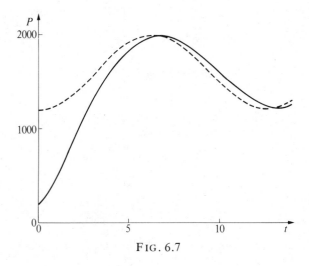

FIG. 6.7

was drawn from points obtained by using a range of values of T and in each case evaluating the integral by Simpson's Rule. The whole process can be programmed on a computer. Note that after the initial rise the population curve matches the bound curve very well but is a little out of step with it. As environmental conditions change there is a slight lag before the population adapts.

The original differential equation is solved directly by a numerical method as an example in §6.8.

The general reduced equation

$$\frac{dy}{dx} + p(x)y = 0$$

can be solved by separating the variables.

$$\int \frac{dy}{y} = -\int p(x)dx + C,$$

$$y = e^{-\int p(x)dx + C} = Ae^{-\int p(x)dx}.$$

In particular, if the coefficient $p(x)$ is a constant p, then the solution of the reduced equation, the complementary function, is a simple exponential,

$$y = Ae^{-px}.$$

Exercises

1. Solve the following first-order linear ordinary differential equations by using an integrating factor.

(a) $\dfrac{dy}{dx} + ay = 0$;

(b) $\dfrac{dy}{dx} + 2y = 3$;

(c) $\dfrac{dy}{dx} + 2xy = 3x$;

(d) $\dfrac{dy}{dx} + \dfrac{2}{x}y = 3$;

(e) $\dfrac{dy}{dx} + \dfrac{2x}{x^2 - 1}y = e^x$;

(f) $\dfrac{dy}{dx} + y \tan x = x^2 e^x \cos x$.

2. Solve the first two parts of question 1 using separation of variables.

3. For each solution of question 1 state which part is the complementary function and which is the particular integral.

4. Show that Bernoulli's Equation

$$\frac{dy}{dx} + p(x)y = q(x)y^n$$

can be converted to linear form by the substitution $v = 1/y^{n-1}$. Solve the equation

$$\frac{dy}{dx} + y = -xy^4, \ y = 1 \text{ when } x = 0.$$

5. A gas at $20\,^\circ\mathrm{C}$ is blown past a body initially at $100\,^\circ\mathrm{C}$ in order to cool it. Under these conditions of forced convection the body cools to $70\,^\circ\mathrm{C}$ in 10 minutes. Assuming Newton's Law of Cooling (§6.1) formulate the differential equation governing this situation. Solve the equation using an integrating factor and hence find the temperature of the body as a function of time.

6. A gas initially at $20\,^\circ\mathrm{C}$ is blown past a body initially at $100\,^\circ\mathrm{C}$ in order to cool it. The gas is recirculated and its temperature T_g at any time t is given by

$$T_g = 30 - 10e^{-\frac{t}{10}}.$$

The temperature T of the body is governed by the differential equation

$$\frac{dT}{dt} = -\frac{1}{20}(T - T_g).$$

Find the temperature of the body as a function of time and determine its final value. How long does it take to get within 5 per cent of this final value?

6.6. Second-order linear

The general *second-order linear* ordinary differential equation has the form

$$a(x)\frac{d^2y}{dx^2} + b(x)\frac{dy}{dx} + c(x)y = f(x),$$

where $a(x)$, $b(x)$, $c(x)$, and $f(x)$ are functions of x. As may be imagined such equations can be very complex and so we will restrict consideration to those with *constant coefficients* a, b, and c. Without loss of generality it can be assumed that $a = 1$ since if it is not the equation can be divided through by it to give a new equation with the coefficient of d^2y/dx^2 equal to unity. We therefore consider an equation

$$\frac{d^2y}{dx^2} + b\frac{dy}{dx} + cy = f(x).$$

To simplify the situation further consider the *reduced equation*

$$\frac{d^2y}{dx^2} + b\frac{dy}{dx} + cy = 0,$$

because if this cannot be solved there is not much hope for the complete equation.

It was noted at the end of the previous section that, if the first-order linear equation had constant coefficients, the solution of the reduced equation, the complementary function, took the form of a simple exponential. The same might be true for the second-order linear equation so we try a solution

$$y = e^{\alpha x}, \alpha \text{ a constant.}$$

Then,
$$\frac{dy}{dx} = \alpha e^{\alpha x}$$

$$\frac{d^2y}{dx^2} = \alpha^2 e^{\alpha x}$$

so that on substitution in the reduced equation we have

$$\alpha^2 e^{\alpha x} + b\alpha e^{\alpha x} + ce^{\alpha x} = 0$$

or

$$\alpha^2 + b\alpha + c = 0.$$

This is known as the *auxiliary equation* and is a quadratic in α showing that in general it will have two solutions, α_1 and α_2. Hence, $y = e^{\alpha x}$, with $\alpha = \alpha_1$ or α_2 is indeed a solution of the reduced equation. We therefore have *two* complementary functions in general,

$$y_{c_1} = e^{\alpha_1 x}, \quad y_{c_2} = e^{\alpha_2 x}$$

where α_1 and α_2 are the two distinct solutions of the auxiliary equation.

A more general solution of the reduced equation is obtained by forming a *linear combination* of the two complementary functions. A linear combination is an expression of the form

$$y_c = Ae^{\alpha_1 x} + Be^{\alpha_2 x}$$

where A and B are constants. The fact that this is a solution can be shown by direct substitution. Note that if $A = 1, B = 0$ the first solution is produced and if $A = 0, B = 1$, the second.

Example. Solve $y'' - y' - 2y = 0$.

The auxiliary equation is $\alpha^2 - \alpha - 2 = 0$ and has solutions $\alpha = -1$ or 2. The two solutions are therefore e^{-x} and e^{2x} which can be linearly combined to give

$$y = Ae^{-x} + Be^{2x}.$$

Example. Solve $y'' + 4y = 0$.

The auxiliary equation is $\alpha^2 + 4 = 0$ and so $\alpha = \pm 2i$ (Appendix A.3),

$$y = Ae^{2i} + Be^{-2i}.$$

These imaginary exponentials may be expressed as sines and cosines (§4.7) so that an alternative more useful form is

$$y = A \cos 2x + B \sin 2x.$$

If the roots of the auxiliary equation are equal only a single complementary function is obtained. This means that if a second complementary function does exist it is not a simple exponential. A more general approach is required.

The auxiliary equation has two roots α_1 and α_2 (which may be equal) and so we can write

$$\alpha^2 + b\alpha + c = (\alpha - \alpha_1)(\alpha - \alpha_2) = 0.$$

If we compare the original reduced equation

$$\frac{d^2 y}{dx^2} + b\frac{dy}{dx} + cy = 0$$

with this we note that it has a 'squared' first term. Two differential operators d/dx are applied to y and so the first term could formally be written

$$\left(\frac{d}{dx}\right)\left(\frac{d}{dx}\right)y.$$

It is important that these differential operators are kept to the left of y. The order of the 'factors' cannot be interchanged. Using (d/dx) as a rather special factor the reduced equation can be written

$$\left(\frac{d}{dx}\right)^2 y + b\left(\frac{d}{dx}\right)y + cy = 0.$$

Comparing this with the auxiliary equation above shows that the reduced equation can be 'factorized' to give

$$\left(\frac{d}{dx} - \alpha_1\right)\left(\frac{d}{dx} - \alpha_2\right)y = 0.$$

This can be multiplied out making sure that the differential operators remain to the left of y to give

$$\left\{\left(\frac{d}{dx}\right)^2 - (\alpha_1 + \alpha_2)\left(\frac{d}{dx}\right) + \alpha_1\alpha_2\right\}y = 0$$

or

$$\left(\frac{d}{dx}\right)^2 y - (\alpha_1 + \alpha_2)\left(\frac{d}{dx}\right)y + \alpha_1\alpha_2 y = 0.$$

This reproduces the reduced equation because $-(\alpha_1 + \alpha_2) = b$ and $\alpha_1\alpha_2 = c$.

Using the 'factorized' form of the reduced equation let

$$\left(\frac{d}{dx} - \alpha_2\right)y = z, \text{ a new variable.}$$

The reduced equation can then be written

$$\left(\frac{d}{dx} - \alpha_1\right)z = 0.$$

This process has split the second-order linear equation

$$\frac{d^2 y}{dx^2} + b\frac{dy}{dx} + cy = 0$$

into a pair of first-order linear equations

$$\frac{dz}{dx} - \alpha_1 z = 0,$$

$$\frac{dy}{dx} - \alpha_2 y = z$$

which can be solved by the methods of §6.5.

The solutions are

$$z = C_1 e^{\alpha_1 x} \quad \text{and} \quad y = e^{\alpha_2 x} \int z e^{-\alpha_2 x} dx + C_2 e^{\alpha_2 x}$$

where C_1 and C_2 are constants. Hence

$$y = e^{\alpha_2 x} \int C_1 e^{(\alpha_1 - \alpha_2)x} dx + C_2 e^{\alpha_2 x}$$

and if $\alpha_1 \neq \alpha_2$ the exponential $e^{(\alpha_1 - \alpha_2)x}$ can be integrated to give

$$y = \frac{C_1}{\alpha_1 - \alpha_2} e^{\alpha_1 x} + C_2 e^{\alpha_2 x}.$$

This has reproduced the pair of complementary functions $e^{\alpha_1 x}$ and $e^{\alpha_2 x}$ found previously. If, however, $\alpha_1 = \alpha_2 = \alpha_0$, say, $e^{(\alpha_1 - \alpha_2)x} = 1$ and so

$$y = e^{\alpha_0 x} C_1 x + C_2 e^{\alpha_0 x} = (C_1 x + C_2) e^{\alpha_0 x}.$$

In this case the pair of C.F.s are $e^{\alpha_0 x}$ and $x e^{\alpha_0 x}$.

The above general method shows that two C.F.s always *exist*.

Example. Solve $y'' - 6y' + 9y = 0$.

Auxiliary equation: $\alpha^2 - 6\alpha + 9 = (\alpha - 3)^2 = 0$.

Hence a double root, $\alpha = 3$. One C.F. is therefore e^{3x} and the other simply $x e^{3x}$. The general solution is a linear combination of these two,

$$y = (A + Bx) e^{3x}.$$

It has been shown that the general solution of the reduced equation can always be found. The general method used can be employed to obtain the solution of the complete equation by splitting it into a pair of first-order linear equations. The equation

$$\frac{d^2y}{dx^2} + b\frac{dy}{dx} + cy = f(x)$$

may be written

$$\left(\frac{d}{dx} - \alpha_1\right)\left(\frac{d}{dx} - \alpha_2\right)y = f(x)$$

which gives the pair

$$\frac{dz}{dx} - \alpha_1 z = f(x),$$

$$\frac{dy}{dx} - \alpha_2 y = z.$$

Solutions are

$$z = e^{\alpha_1 x}\int f(x)e^{-\alpha_1 x}dx + C_1 e^{\alpha_1 x},$$

$$y = e^{\alpha_2 x}\int z e^{-\alpha_2 x}dx + C_2 e^{\alpha_2 x},$$

from which

$$y = \underbrace{Ae^{\alpha_1 x} + Be^{\alpha_2 x}}_{\text{C.F.}} + \underbrace{e^{\alpha_2 x}\int e^{(\alpha_1 - \alpha_2)x}\int^x f(u)e^{-\alpha_1 u}dudx}_{\text{P.I.}}$$

if $\alpha_1 \neq \alpha_2$ and

$$y = \underbrace{(A + Bx)e^{\alpha_0 x}}_{\text{C.F.}} + \underbrace{e^{\alpha_0 x}\iint^x f(u)e^{-\alpha_0 u}dudx}_{\text{P.I.}}$$

if there is a double root, α_0, of the auxiliary equation.

The particular integral can be obtained from one or other of the general integrals above. This can be rather a complicated process and for the simpler functions shorter methods are known or can be devised. These methods rely on the fact that a second-order linear ordinary differential equation has a unique solution. Any process can therefore be used to try and obtain a solution with the knowledge that if the solution so obtained satisfies the equation it must be the one and only correct solution.

If the right-hand side, $f(x)$, of the differential equation

$$\frac{d^2y}{dx^2} + b\frac{dy}{dx} + cy = f(x)$$

is a polynomial we require as a particular solution a function, which, when substituted in the left-hand side, will generate a polynomial identical to $f(x)$ so that the equation is satisfied. A little thought shows that if a polynomial is substituted in the left-hand side a polynomial is generated. Furthermore, provided $c \neq 0$, the polynomial generated will be of the same degree as that of the polynomial substituted. If $c = 0$ and $b \neq 0$ the polynomial generated will be of degree one less than that substituted and if both b and c are zero it will be of degree two less. With these thoughts in mind a reasoned guess can be made at the general form of the particular integral which can then be tried and adjusted to satisfy the equation.

Example. Find the particular integral of $y'' - 3y' + 2y = x^2 + 2$.

The coefficient of y is non-zero and so the polynomial generated will be of the same degree as that substituted in. In order to satisfy the equation a quadratic must be generated and so a general quadratic is tried as a particular integral.

Try $y = ax^2 + bx + c$ so that $y' = 2ax + b$ and $y'' = 2a$.

Then

$$y'' - 3y' + 2y = 2ax^2 + (2b - 6a)x + 2a - 3b + 2c.$$

In order to satisfy the equation this expression must be identical to $x^2 + 2$. Comparing coefficients shows that $2a = 1$, $2b - 6a = 0$, and $2a - 3b + 2c = 2$. Hence

$$a = \frac{1}{2}, \quad b = \frac{3}{2}, \quad c = \frac{11}{4}$$

and the particular integral is

$$y = \frac{1}{2}x^2 + \frac{3}{2}x + \frac{11}{4}.$$

Example. Find the particular integral of $y'' + y' = 3$.

There is no term in y on the left-hand side and so the polynomial generated will be of degree one less than that substituted. The right-hand side is of degree zero and so a polynomial of degree one is tried for the particular integral.

Try $y = ax + b$ so that $y' = a$ and $y'' = 0$. Note that b is differentiated out. Substitution gives $y'' + y' = a$ and in order to satisfy the equation we must choose $a = 3$. The P.I. appears to be $y = 3x + b$ where b is an arbitrary constant. The reason for this becomes apparent if the C.F.s are found. The auxiliary equation is $\alpha^2 + \alpha = 0$ and so the C.F.s are 1 and e^{-x}. The former is a constant and so the constant b is really a C.F. and the trial P.I. should have been $y = ax$ resulting in $y = 3x$ as the true P.I. For this reason it is always advisable to find the C.F.s and then ensure that they are not included in any trial form of the P.I.

Example. A certain environment can support a maximum population P_M of a given species. Initially the population is $\frac{1}{5}P_M$ and is growing at a rate of $\frac{1}{20}P_M$ per month. A population explosion occurs which upsets the environmental balance and the population exceeds P_M. Because the environment cannot support the population it subsequently falls to below P_M and decreasing oscillations about P_M occur. Such a population explosion may be described by the differential equation

$$\frac{\mathrm{d}^2 P}{\mathrm{d}t^2} + \frac{\mathrm{d}P}{\mathrm{d}t} + \frac{5}{4}P = \frac{5}{4}P_M,$$

where P is the size of the population and t is the time in months. Find an expression for P as a function of t and find the largest value attained by P and the time at which this occurs.

This is a standard second-order linear ordinary differential equation with a constant right hand side. The auxiliary equation is

$$\alpha^2 + \alpha + \frac{5}{4} = 0$$

which has solutions

$$\alpha = \frac{-1 \pm \sqrt{(1 - 5)}}{2} = -\frac{1}{2} \pm i.$$

The C.F.s are therefore $e^{-\frac{1}{2}t}\cos t$ and $e^{-\frac{1}{2}t}\sin t$. The right-hand-side is a constant and so we may expect the P.I. to be a constant. A trial shows that the P.I. is P_M and so the general solution is

$$P = (A \cos t + B \sin t)e^{-\frac{1}{2}t} + P_M.$$

The initial conditions are given as $P = \frac{1}{5}P_M$ and $\mathrm{d}P/\mathrm{d}t = \frac{1}{20}P_M$. Using

$$\frac{\mathrm{d}P}{\mathrm{d}t} = (A \cos t + B \sin t) \left(-\frac{1}{2}\right) e^{-\frac{1}{2}t} + (-A \sin t + B \cos t)e^{-\frac{1}{2}t}$$

and substituting in the initial conditions $(t = 0)$ shows that

$$A = -\frac{4}{5}P_M \quad \text{and} \quad B = -\frac{7}{20}P_M.$$

Hence

$$P = P_M \left\{ 1 - \left(\frac{4}{5} \cos t + \frac{7}{20} \sin t\right) e^{-\frac{1}{2}t} \right\}.$$

Note that the oscillatory part of the solution decays with time due to the negative exponential and so the population tends towards P_M. A graph of the solution is shown in Figure 6.8.

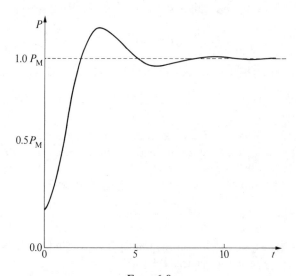

FIG. 6.8

For critical points, we require to solve

$$\frac{dP}{dt} = P_{M} \left(\frac{1}{20} \cos t + \frac{39}{40} \sin t \right) e^{-\frac{1}{2}t} = 0.$$

The solutions are $\tan t = -(2/39)$ or $t = \infty$. The latter is the steady state population after the oscillations have died away. The former yields $t = n\pi - 0.05124$ rad (or $180n - 2.9357°$) where n is any integer. The first positive solution is $t = 3.0904$, which, from the graph (Figure 6.8) is a maximum. The second derivative is

$$\frac{d^2P}{dt^2} = P_{M} \left\{ \frac{19}{20} \cos t - \frac{43}{80} \sin t \right\} e^{-\frac{1}{2}t}$$

which is negative when $t = 3.0904$. A maximum at this value of t is therefore confirmed. The value of the maximum is $P = 1.16657P_{M}$. There are other maxima because the solution oscillates but the exponential damping ensures that they decrease in magnitude and so the first is the largest.

If the right-hand side, $f(x)$, of the equation is an exponential the P.I. must be such that an identical exponential is generated from the left-hand side. The trial P.I. must therefore contain an exponential with an exponent identical to that of the right-hand side. If $f(x) = Ae^{\alpha x}$ try $y = Ce^{\alpha x}$ as a P.I. Substitution shows that the left-hand side is

$$\frac{d^2y}{dx^2} + b\frac{dy}{dx} + cy = C(\alpha^2 + b\alpha + c)e^{\alpha x}.$$

In order to satisfy the equation this must be identical to $Ae^{\alpha x}$ and so

$$C = \frac{A}{\alpha^2 + b\alpha + c}$$

provided $\alpha^2 + b\alpha + c \neq 0$. That is provided the assumed form of the P.I. is not the same as either of the C.F.s. If $Ce^{\alpha x}$ is a C.F. try $Cxe^{\alpha x}$ as a P.I. If this also is a C.F. try $Cx^2e^{\alpha x}$. These forms can be obtained by evaluating the general integrals for the P.I. derived earlier.

Example. Solve $y'' + 3y' + 2y = 3e^{-x}$.

C.F.s are e^{-x} and e^{-2x}. The right-hand side is therefore an exponential of C.F. type and so the trial P.I. used is

$$y = Cxe^{-x}.$$

Then

$$y' = -Cxe^{-x} + Ce^{-x} = C(1-x)e^{-x},$$

$$y'' = -C(1-x)e^{-x} - Ce^{-x} = -C(2-x)e^{-x}.$$

Substitution gives

$$-C(2-x)e^{-x} + 3C(1-x)e^{-x} + 2Cxe^{-x} = 3e^{-x}.$$

Therefore, $$C = 3,$$

$$y = Ae^{-x} + Be^{-2x} + 3xe^{-x}.$$

If $f(x) = A \sin \omega x$, a trigonometric function, the trial P.I. can contain both $\sin \omega x$ and $\cos \omega x$ since both can give rise to $\sin \omega x$ terms due to the derivatives on the left-hand side. A linear combination

$$y = C_1 \sin \omega x + C_2 \cos \omega x$$

is the most general form. If the C.F. is of this form try

$$y = x(C_1 \sin \omega x + C_2 \cos \omega x).$$

Example. Solve $y'' - y = 3 \cos 2x + 2 \sin 2x$.

C.F.s are e^x and e^{-x}. Try $y = C_1 \sin 2x + C_2 \cos 2x$ as a P.I.

Then

$$y' = 2C_1 \cos 2x - 2C_2 \sin 2x,$$

$$y'' = -4C_1 \sin 2x - 4C_2 \cos 2x.$$

Substitution gives

$$-4C_1 \sin 2x - 4C_2 \cos 2x - C_1 \sin 2x + C_2 \cos 2x = 3 \cos 2x + 2 \sin 2x.$$

Comparing the coefficients of $\sin 2x$ and $\cos 2x$ on both sides shows that

$$C_1 = -\frac{2}{5}, C_2 = -\frac{3}{5}.$$

The general solution is

$$y = Ae^x + Be^{-x} - \frac{2}{5} \sin 2x - \frac{3}{5} \cos 2x.$$

Example. Solve $y'' + y = \sin x$ with the initial conditions that

$$y = 1 \text{ and } y' = \frac{1}{2} \text{ when } x = 0.$$

The reduced equation is $y'' + y = 0$ and so the C.F.s are $\sin x$ and $\cos x$. The general solution of the reduced equation is therefore $y = A \sin x + B \cos x$.

In this case the right-hand side of the complete equation is $\sin x$, which is the form of one of the C.F.s. We therefore use as a trial P.I.

$$y = x(C_1 \sin x + C_2 \cos x).$$

Then

$$y' = x(C_1 \cos x - C_2 \sin x) + (C_1 \sin x + C_2 \cos x)$$

and

$$y'' = x(-C_1 \sin x - C_2 \cos x) + 2(C_1 \cos x - C_2 \sin x).$$

Hence

$$y'' + y = x(-C_1 \sin x - C_2 \cos x) + 2(C_1 \cos x - C_2 \sin x)$$
$$+ x(C_1 \sin x + C_2 \cos x)$$
$$= 2(C_1 \cos x - C_2 \sin x).$$

If this is to be identically equal to $\sin x$ then $C_1 = 0$ and $C_2 = -\frac{1}{2}$. The general solution of the equation is therefore

$$y = A \sin x + B \cos x - \frac{1}{2} x \cos x.$$

Since the initial conditions of both y and y' have been given it is possible to find the constants A and B. Since $y = 1$ when $x = 0$, $B = 1$. Also

$$y' = A \cos x - B \sin x + \frac{1}{2} x \sin x - \frac{1}{2} \cos x$$

and if this equals $\frac{1}{2}$ when $x = 0$ then $A = 1$. The solution which satisfies the differential equation with the given initial conditions is therefore

$$y = \cos x + \sin x - \frac{1}{2} x \cos x.$$

Example. A population P exists in an environment which shows seasonal variations in its ability to support the population. An external influence triggers a population explosion which is described by the differential equation

$$\frac{d^2P}{dt^2} + \frac{dP}{dt} + \frac{5}{4}P = P_M\left\{\frac{5}{4} - \frac{2}{5}\cos\frac{1}{2}t + \frac{1}{5}\sin\frac{1}{2}t\right\}$$

where t is the time in months. The constant P_M is the mean value of the maximum population capable of being supported by the environment. If the population is initially of size $\frac{2}{3}P_M$ and at this time is increasing at the rate of $\frac{8}{5}P_M$ per month find P as a function of t.

The auxiliary equation is $\alpha^2 + \alpha + \frac{5}{4} = 0$ which has solutions $\alpha = -\frac{1}{2} \pm i$. The C.F.s are therefore

$$e^{-\frac{1}{2}t}\cos t \text{ and } e^{-\frac{1}{2}t}\sin t.$$

For the P.I. try

$$P = P_M\left\{a + b\cos\frac{1}{2}t + c\sin\frac{1}{2}t\right\}$$

since the right-hand side of the differential equation is the sum of a constant, $\frac{5}{4}$, and a linear combination of $\cos\frac{1}{2}t$ and $\sin\frac{1}{2}t$. Then

$$\frac{dP}{dt} = P_M\left\{-\frac{1}{2}b\sin\frac{1}{2}t + \frac{1}{2}c\cos\frac{1}{2}t\right\}$$

and

$$\frac{d^2P}{dt^2} = P_M\left\{-\frac{1}{4}b\cos\frac{1}{2}t + \frac{1}{4}c\sin\frac{1}{2}t\right\}.$$

Substituting these into the left-hand-side of the differential equation gives

$$\frac{d^2P}{dt^2} + \frac{dP}{dt} + \frac{5}{4}P = P_M\left\{\frac{5}{4}a + \left(b + \frac{1}{2}c\right)\cos\frac{1}{2}t + \left(c - \frac{1}{2}b\right)\sin\frac{1}{2}t\right\}.$$

This must be identical to the original right-hand-side and so $a = 1$, $b = -\frac{2}{5}$, and $c = 0$. The general solution is therefore

$$P = P_M\left\{(A\cos t + B\sin t)e^{-\frac{1}{2}t} + 1 - \frac{2}{5}\cos\frac{1}{2}t\right\}.$$

Initially $P = \frac{2}{3}P_M$ and so $A = -\frac{1}{5}$. Also $dP/dt = \frac{8}{5}P_M$ when $t = 0$. Using

$$\frac{dP}{dt} = P_M \left\{ \left(B - \frac{1}{2}A \right) \cos t - \left(A + \frac{1}{2}B \; \sin t \right) e^{-\frac{1}{2}t} + \frac{1}{5} \sin \frac{1}{2}t \right\}$$

shows that $B = \frac{3}{2}$. Hence

$$P = P_M \left\{ \left(\frac{3}{2} \sin t - \frac{1}{5} \cos t \right) e^{-\frac{1}{2}t} + 1 - \frac{2}{5} \cos \frac{1}{2}t \right\}.$$

As t increases P tends towards $P_M \{1 - \frac{2}{5} \cos \frac{1}{2}t\}$. The initial explosion is exponentially damped out and a periodic steady state ensues. A graph of P against t is shown in Figure 6.9. The dotted line is the steady-state function.

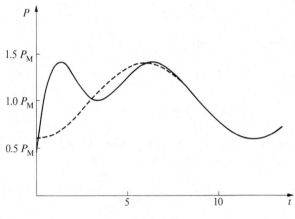

FIG. 6.9

Exercises

1. Solve the following differential equations.

(a) $y'' - 4y = 0$;   (b) $y'' + 9y = 0$;

(c) $y'' - 3y' + 2y = 0$;   (d) $y'' - 2y' + y = 0$.

2. Solve the following differential equations with the given initial conditions.

(a) $y'' - y = 0$, $y = 2$ and $y' = 0$ when $x = 0$;

(b) $y'' - 5y' + 6y = 0$, $y = 1$ and $y' = 1$ when $x = 0$;

(c) $y'' + 4y = 0$, $y = 2$ and $y' = -2$ when $x = 0$;

(d) $y'' + 4y' + 4y = 0$, $y = 1$ and $y' = 3$ when $x = 0$.

3. Find the C.F.s and P.I.s of the following differential equations and hence the general solutions. In addition evaluate the arbitrary constants where initial conditions are given.

(a) $y'' - 9y = 2x$;

(b) $y'' + y' + y = x^2$;

(c) $y'' - 2y' - 3y = e^{2x}$, $y = \dfrac{2}{3}, y' = \dfrac{1}{3}$ when $x = 0$;

(d) $y'' - y' - 2y = e^{2x}$;

(e) $y'' - 2y' + y = e^x$;

(f) $y'' + y' - 6y = 3\cos 2x$, $y = \dfrac{37}{52}$, $y' = -\dfrac{46}{52}$ when $x = 0$;

(g) $y'' + 2y' - 3y = 2\sin 3x - \cos 3x$;

(h) $y'' + 4y = \sin 2x$, $y = 1, y' = \dfrac{3}{4}$ when $x = 0$.

4. A cylindrical hydrometer of length 10 cm is floating with its axis vertical in a non-viscous liquid. When at rest the hydrometer floats with 2 cm above the surface. If the hydrometer is pushed vertically downwards until only 1 cm is above the surface and then released its subsequent motion is governed by the differential equation

$$\frac{d^2x}{dt^2} + \frac{x}{8} = 0, \quad x = 1 \text{ and } \frac{dx}{dt} = 0 \text{ when } t = 0,$$

where x is the displacement from the equilibrium position. Find x as a function of time. Derive the differential equation using Archimedes' Principle.

5. If the hydrometer of question 4 were floated in a viscous liquid an extra term would appear in the differential equation to allow for the viscous damping of motion. The equation would then have the form

$$\frac{d^2x}{dt^2} + 2k\frac{dx}{dt} + \frac{x}{8} = 0, \quad x = 1, \frac{dx}{dt} = 0 \text{ when } t = 0,$$

where k is a constant depending upon the viscosity. Find x as a function of time for the three cases $k^2 < \frac{1}{8}$, $k^2 = \frac{1}{8}$, and $k^2 > \frac{1}{8}$. Sketch the graph of x against t in each case.

6.7. Simultaneous differential equations

Problems can arise in which the time rate of change of a dependent variable is a function of another dependent variable. In such cases it may be possible to formulate a second relationship expressing the time rate of change of the second dependent variable. If this can be done the result is a pair of *simultaneous differential equations* from which it may be hoped to find each of the dependent variables as functions of time. Examples of this situation are found in the relationship between predator and prey, in populations competing with each other in a given environment, and in related chemical reactions.

If the equations are sufficiently simple it may be possible to eliminate one of the dependent variables and its derivatives and so obtain a single differential equation for the other dependent variable.

Example. Solve the pair of simultaneous differential equations

$$\frac{dx}{dt} = \frac{y}{5} - 2 \quad \text{and} \quad \frac{dy}{dt} = 10 - 5x$$

where initially $x = 2$ and $y = 5$.

The dependent variable y may be eliminated by differentiating the first equation and substituting for dy/dt from the second.

$$\frac{d^2x}{dt^2} = \frac{1}{5}\frac{dy}{dt} = \frac{1}{5}(10 - 5x) = 2 - x.$$

The equation in x is therefore

$$\frac{d^2x}{dt^2} + x = 2$$

which has C.F.s $\sin t$ and $\cos t$ and P.I. 2. Therefore,

$$x = A \sin t + B \cos t + 2.$$

Now $x = 2$ when $t = 0$ so that $B = 0$. From the first equation $dx/dt = -1$ when $t = 0$. From the general solution,

$$\frac{dx}{dt} = A \cos t - B \sin t$$

and so $A = -1$. Hence

$$x = 2 - \sin t$$

and again from the first equation

$$y = 5(dx/dt + 2) = 5(-\cos t + 2) = 10 - 5 \cos t.$$

Example. In a chemical decomposition a compound X produces a compound Y which in turn decomposes to give a compound Z. The decompositions are governed by the equations

$$\frac{dx}{dt} = -4x, \quad \frac{dy}{dt} = 3x - 2y, \quad \frac{dz}{dt} = y,$$

where x, y, and z are the masses of X, Y, and Z, respectively, and time is measured in hours. If initially there are 8g of X and no Y or Z, find x, y, and z as functions of time and hence determine the maximum value of y and the final value of z.

Note that the first equation is separable and can be integrated to give

$$x = 8e^{-4t}$$

after using the initial condition on x. The second equation can now be written

$$\frac{dy}{dt} + 2y = 24e^{-4t}$$

which is linear. The integrating factor is e^{2t} and so

$$\frac{d}{dt}(ye^{2t}) = 24e^{-2t}.$$

Hence,

$$ye^{2t} = -12e^{-2t} + C_1$$

or

$$y = -12e^{-4t} + C_1e^{-2t}.$$

Since $y = 0$ when $t = 0$, the value of C_1 is 12. Therefore,

$$y = 12(e^{-2t} - e^{-4t}).$$

The third equation can now be solved by direct integration since

$$\frac{dz}{dt} = 12(e^{-2t} - e^{-4t}).$$

Therefore, $$z = -6e^{-2t} + 3e^{-4t} + C_2,$$

and $C_2 = 3$ if $z = 0$ at $t = 0$.

The masses x, y, and z are therefore given by

$$x = 8e^{-4t}, \quad y = 12(e^{-2t} - e^{-4t}), \quad z = 3(1 + e^{-4t} - 2e^{-2t}).$$

From this it is clear that the final value of z is 3g. The maximum value of y occurs when

$$\frac{dy}{dt} = 12(-2e^{-2t} + 4e^{-4t}) = 0.$$

That is when

$$e^{-2t} = \frac{1}{2} \text{ or } t = 0.347 \text{ hours.}$$

This must in fact be a maximum since both the initial and final values of y are zero. Substituting for e^{-2t} shows that

$$y_{max} = 3g.$$

This example, which is illustrated in Figure 6.10, shows that what may at first appear to be a rather involved set of simultaneous differential equations can in practice be relatively simple because of the form the equations take. Such direct methods of approach should be considered before embarking on a lengthy elimination of variables.

Exercises

1. Solve the following sets of simultaneous differential equations.

(a) $\dfrac{dx}{dt} = -2x, \dfrac{dy}{dt} = x - 2y, x = 5, y = 2$ when $t = 0$;

(b) $\dfrac{dx}{dt} = 3 - 3x + y, \dfrac{dy}{dt} = 2 - 2x, x = 1, y = 1$ when $t = 0$;

(c) $\dfrac{dx}{dt} = x - y + 1, \dfrac{dy}{dt} = 5x - 3y + 3, x = 0, y = 0$ when $t = 0$;

(d) $3\dfrac{dx}{dt} = 1 - 7x + 2y, 3\dfrac{dy}{dt} = 11 + 4x - 5y, x = 1, y = 0$ when $t = 0$.

2. In a predator–prey situation the rate of growth of the predator

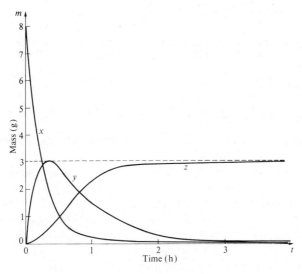

FIG. 6.10

population is proportional to the size of the prey population. The rate of change of the prey population however depends upon the size of both predator and prey populations. The differential equations describing such a relationship are

$$100\frac{dx}{dt} = y, \quad 100\frac{dy}{dt} = 2y - x$$

where x and y are the numbers of predator and prey respectively and t is the time in years. If initially the predator population is 10 thousand and the prey population is 5 million, find each as a function of time. How large is each population in 5 years time?

3. Two different concentrations of a solution are separated by a membrane through which the solute can diffuse. The rate at which the solute diffuses is proportional to the difference in concentrations between the two solutions and the direction is such as to equalize these concentrations. The differential equations governing the process are

$$\frac{dC_1}{dt} = -\frac{k}{V_1}(C_1 - C_2) \quad \text{and} \quad \frac{dC_2}{dt} = \frac{k}{V_2}(C_1 - C_2)$$

where C_1 and C_2 are the two concentrations, V_1 and V_2 are the volumes of the respective compartments containing these concentrations, and k is a constant of proportionality. If $V_1 = 20$ litres, $V_2 = 5$ litres, and $k = 0.2$ litres/min and if initially $C_1 = 3$ moles/litre and $C_2 = 0$, find C_1 and C_2 as functions of time.

6.8. Numerical methods

The previous sections have dealt with analytical methods of solving the more commonly occurring types of ordinary differential equations. Further analytical techniques are available for more complicated equations and can be found in specialist books. Even so equations occur for which no analytical method of solution is known or whose analytical solution is so complicated that it is of little use. In such cases *numerical methods* may be employed to try and calculate the values taken by the dependent variable over a range of values of the independent variable.

Numerical methods are approximate methods, but, with care, the approximations can be made so good that values calculated from the analytical solution and by a numerical method are virtually identical. These advanced numerical methods are themselves complicated but do yield results which cannot be obtained in any other way. However, the principles involved can be illustrated using more elementary methods. If, for example, a first order equation is to be solved and a starting point is known then the slope of the solution curve at this starting point can be calculated from the differential equation itself. Hence a point and the slope of the solution curve at this point are known. A step can now be taken to a neighbouring solution point by assuming that the slope of the solution curve does not change significantly between the two points. The direction in which this step is to be taken is therefore known. Provided that the step is not too large this assumption is reasonable. At the neighbouring solution point the slope is recalculated from the differential equation which gives the direction of the next step to be taken. Progress is therefore made step by step. A knowledge of the values of the dependent variable and its derivative(s) at one point is used to calculate these values at a neighbouring point. These values are then sometimes used to improve the calculation.

Assume that the equation to be 'solved' is the general first-order equation

$$y' = f(x, y)$$

and that the point (x_0, y_0) is known. The fact that when $x = x_0$, $y = y_0$ means that had it been possible to solve the differential equation analytically the single arbitrary constant which would have arisen could have been evaluated. A specific solution curve is obtained. In a numerical method a starting point must be known from which a specific solution curve is constructed. The starting point here is (x_0, y_0) and the original differential equation shows that the slope at this point is

$$y_0' = f(x_0, y_0).$$

We wish to calculate y at a neighbouring point to x. Let this point be $x_0 + h$, where h is known as the *step length*, and denote it by x_1 so that

$$x_1 = x_0 + h.$$

Now the solution y is a function of the independent variable x, that is

$$y = y(x)$$

and in particular $y_0 = y(x_0)$. Using Taylor's Theorem (§4.7)

$$y_1 = y(x_1) = y(x_0 + h) = y(x_0) + hy'(x_0) + \frac{h^2}{2!}y''(x_0) + \ldots$$

$$= y_0 + hy_0' + \frac{h^2}{2!}y_0'' + \ldots$$

If it is assumed that h is chosen sufficiently small for terms involving h^2 and higher powers of h to be neglected then

$$y_1 = y_0 + hy_0'$$

and

$$y_0' = f(x_0, y_0).$$

The value of y_1 can therefore be expressed in terms of x_0 and y_0. Using the same process (x_1, y_1) can be used to calculate the next point (x_2, y_2) and so on. In this way *recurrence relationships* are set up from which (x_{r+1}, y_{r+1}) can be calculated if (x_r, y_r) is known.

$$y_r' = f(x_r, y_r),$$

$$x_{r+1} = x_r + h,$$

$$y_{r+1} = y_r + hy_r'.$$

This simplest of processes is known as *Euler's method*.

Geometrically the solution curve is approximated to by a series of straight lines. The horizontal projection of each of these lines is of length h and their slopes are calculated from the original differential equation. From a known starting point, (x_0, y_0), a line of slope $y_0' = f(x_0, y_0)$ is drawn until it meets the vertical line $x = x_1 = x_0 + h$. This gives the point (x_1, y_1). A new line starting at (x_1, y_1) and of slope $y_1' = f(x_1, y_1)$ is then drawn in the same manner to intercept the line $x = x_2 = x_1 + h$. Continuing in this way gives a straight-line sectional approximation to the solution curve (Figure 6.11).

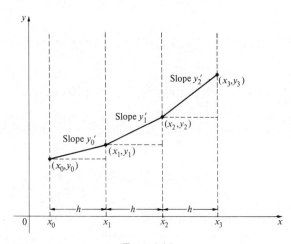

FIG. 6.11

Example. Solve the differential equation

$$\frac{dy}{dx} = x^2 - \frac{y}{x}, \quad y = 1 \text{ when } x = 1,$$

by Euler's method for values of x from 1 to 2 in steps of 0.1.

Comparing with the general method above, $y' = f(x, y) = x^2 - y/x$, $x_0 = 1, y_0 = 1$ and $h = 0.1$.

$$y_r' = x_r^2 - \frac{y_r}{x_r},$$

$$x_{r+1} = x_r + 0.1,$$

$$y_{r+1} = y_r + 0.1y_r'.$$

It is usually most convenient to arrange the calculations in the form of a table. A table also reduces the chances of making mistakes.

r	x_r	y_r	$y_r' = x_r^2 - y_r/x_r$
0	1	1	0
1	1.1	1	0.30091
2	1.2	1.03009	0.58159
3	1.3	1.08825	

In each case the next value of x is obtained by adding 0.1 to the current value and the next value of y by adding one-tenth of the current value of y' to the current value of y.

This repetitive process is particularly suited to the computer which can produce all the required values in milliseconds. A program in BASIC to print out corresponding values of x and y is given below.

```
 10  PRINT "EULER: DY/DX = X ↑ 2 − Y/X"
 20  PRINT
 30  PRINT "X", "Y"
 40  PRINT
 50  X = 1
 60  Y = 1
 70  PRINT X, Y
 80  D = X ↑ 2 − Y/X
 90  X = X + 0.1
100  IF X > 2 THEN 999
110  Y = Y + 0.1 * D
120  GOTO 70
999  END
```

The differential equation has been solved analytically in the first example of §6.5. This solution was obtained without specified values of x and y and so it contains an arbitrary constant. For the values above the solution is

$$y = \frac{1}{4}\left(x^3 + \frac{3}{x}\right).$$

This can be used to compare the numerical solution with the exact analytical solution.

x	y (numerical)	y (exact)	Per cent error
1	1	1	-0
1.1	1	1.01457	-1.4
1.2	1.03009	1.05700	-2.5
1.3	1.08825	1.12617	-3.4
1.4	1.17354	1.22171	-3.9
1.5	1.28571	1.34375	-4.3
1.6	1.42450	1.49275	-4.5
1.7	1.59194	1.66943	-4.6
1.8	1.78729	1.87467	-4.7
1.9	2.01200	2.10949	-4.6
2	2.26711	2.37500	-4.5

The accuracy can be improved by using a shorter step interval h. This however will increase the number of calculations which have to be made to cover the interval from 1 to 2.

Example. During a consideration of bounded population growth (example, §6.5) in which the environmental bound varied periodically the differential equation

$$\frac{dP}{dt} = kP\left\{P_M\left(1 - \frac{1}{4}\cos\frac{1}{2}t\right) - P\right\}$$

with $k = 1/1000$ and $P_M = 1600$ was used. Although an appropriate substitution converted this to linear form an integral arose which could not be evaluated analytically. A solution curve was produced (Figure 6.7) based upon the evaluation of the integral by Simpson's Rule. If the initial size of the population is 200 use Euler's method to find values of P at intervals of 0.5 in t.

The method involves the use of the equations

$$P'_r = P_r\left\{1600\left(1 - \frac{1}{4}\cos\frac{1}{2}t_r\right) - P_r\right\}/1000,$$

$$t_{r+1} = t_r + h,$$

$$P_{r+1} = P_r + hP'_r,$$

where

$$t_0 = 0 \text{ and } P_0 = 200.$$

The step interval h can be 0.5 or some smaller interval provided that the required values of P at 0.5 intervals in t are obtained. Using a step interval of $h = 0.5$ the following table can be constructed.

r	t_r	P_r	P'_r
0	0	200.00	200.00
1	0.5	300.00	273.73
2	1.0	436.87	354.78
3	1.5	614.26	425.72
4	2.0	827.12	460.51
5	2.5	1057.38	440.39
6	3.0	1277.58	375.77
7	3.5	1465.47	301.64
8	4.0	1616.29	242.72
9	4.5	1737.65	197.43
10	5.0	1836.37	154.42
11	5.5	1913.58	

The new values of t_r and P_r are generated from the previous line.

$$t_{11} = t_{10} + 0.5 = 5.0 + 0.5 = 5.5,$$

$$P_{11} = P_{10} + 0.5 \times P'_{10} = 1836.37 + 0.5 \times 154.42 = 1913.58.$$

The next value of the derivative P'_{11} is calculated directly from the original differential equation.

$$P'_{11} = P_{11} \left\{ 1600 \left(1 - \frac{1}{4} \cos\left(\frac{1}{2} \times t_{11} \right) \right) - P_{11} \right\} / 1000$$

$$= 1913.58 \times \left\{ 1600 \left(1 - \frac{1}{4} \cos\left(\frac{1}{2} \times 5.5 \right) \right) - 1913.58 \right\} / 1000$$

$$= 107.43.$$

Calculation continues in this way until the table is complete. Such repetitive calculations are ideal for a computer. The following program in BASIC prints out the population size and the bound at intervals of 0.5 in t. Both these values are rounded to the nearest whole number. The program allows for different initial populations, total times and step lengths to be entered.

```
 10  PRINT: PRINT "EULER FOR P' = P * (1600 * (1 − 0.25 *
     COS(0.5 * T)) − P)/1000": PRINT
 20  DEFFNA(T) = P * (1600 * (1 − 0.25 * COS(0.5 * T)) − P)/1000
 30  T = 0
 40  INPUT "INPUT INITIAL VALUE OF POPULATION P", P: PRINT
 50  INPUT "INPUT TOTAL TIME T0", T0: PRINT
 60  INPUT "INPUT STEP LENGTH H", H: PRINT
 70  PRINT T0/H; "STEPS": PRINT
 80  PRINT "TIME", "POPULATION", "BOUND": PRINT
 90  PRINT T, INT(P + 0.5), INT(1600 * (1 − 0.25 * COS(0.5 * T)) + 0.5)
100  P = P + H * FNA(T): T = T + H: IF T > T0 THEN 120
110  IF 2 * T = INT(2 * T) THEN 90: GOTO 100
120  END
```

This program has multiple statements on some lines, each being separated by a colon. They could have been put on separate lines. In lines 90 and 110 the "integer part of" function INT( ) is used to produce whole numbers from decimals. The addition of 0.5 in INT(P + 0.5) rounds P to the nearest whole number. The use of INT( ) in line 110 is to obtain a print out only at 0.5 intervals of t, starting from $t = 0$. For greater accuracy many intermediate steps could be taken but print out only occurs at these values of t. If a step length of $h = 0.05$ is used then 10 calculations are performed for every one that is printed out.

If the program is run with an initial population of 200, a total time of 15 (months) and a step interval of 0.05 then 300 steps are taken and at each step a new value of P is produced. The only values printed out, however, are those at times t of 0, 0.5, 1.0, 1.5 etc. The complete print out is shown below.

```
RUN
EULER FOR P' = P * (1600 * (1 − 0.25 * COS(0.5 * T)) − P)/1000
INPUT INITIAL VALUE OF POPULATION P? 200
INPUT TOTAL TIME T0? 15
INPUT STEP LENGTH H? 0.05
300 STEPS
```

TIME	POPULATION	BOUND
0	200	1200
.5	319	1212
1	481	1249
1.5	680	1307
2	897	1384
2.5	1110	1474
3	1300	1572

TIME	POPULATION	BOUND
3.5	1465	1671
4	1606	1766
4.5	1726	1851
5	1825	1920
5.5	1902	1970
6	1955	1996
6.5	1984	1998
7	1986	1975
7.5	1964	1928
8	1919	1861
8.5	1856	1778
9	1778	1684
9.5	1691	1585
10	1600	1487
10.5	1511	1395
11	1427	1317
11.5	1353	1256
12	1294	1216
12.5	1252	1200
13	1228	1209
13.5	1226	1243
14	1247	1298
14.5	1289	1373
15	1352	1461

END PROGRAM

A graph of these results is shown in Figure 6.12. The population curve is the continuous line and the bound curve is the dotted line. This graph can be compared with that shown in Figure 6.7 where the same problem was solved using a different method.

A method for solving the general second-order equation

$$y'' = f(x, y, y')$$

can be formulated using the first three terms of the Taylor series

$$y(x + h) = y(x) + hy'(x) + \frac{h^2}{2!}y''(x) + \dots .$$

In this case it is assumed that terms involving h^3 and higher powers may be neglected. The third term is included since it contains the second derivative required for a second-order equation. The general analytical solution of such an equation would have two arbitrary constants. A numerical solution cannot have such constants since it essentially

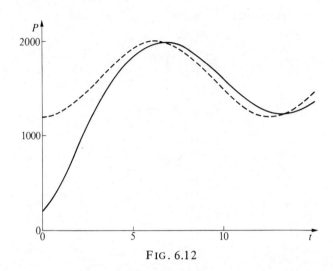

FIG. 6.12

constructs a particular solution curve. Sufficient additional information must therefore be available to eliminate these constants. This is equivalent to a knowledge of a particular point on the solution curve and the slope of the curve at this point, i.e.

$$y = y_0 \text{ and } y' = y'_0 \text{ at } x = x_0$$

where x_0, y_0, and y'_0 are known. Hence, from the original equation

$$y''_0 = f(x_0, y_0, y'_0)$$

and so from the Taylor series

$$y_1 = y_0 + hy'_0 + \frac{1}{2}h^2 y''_0.$$

In addition

$$x_1 = x_0 + h.$$

In order to calculate y_2 from the series y_1, y'_1, and y''_1 must be known. The point (x_1, y_1) has just been calculated and y'_1 can be found from the first two terms of the Taylor series for $y'(x + h)$,

$$y'_1 = y'_0 + hy''_0.$$

Then,

$$y''_1 = f(x_1, y_1, y'_1)$$

and so

$$x_2 = x_1 + h,$$

$$y_2 = y_1 + hy_1' + \frac{1}{2}h^2 y_1''.$$

In this way the next point can be calculated from current values of x, y, y', and y''.

Example. Solve the differential equation

$$\frac{d^2 y}{dx^2} + y = 0, y = 0, \frac{dy}{dx} = 1 \text{ when } x = 0,$$

numerically for values of x from 0 to 1.6 in steps of 0.1.

This is an equation for simple harmonic motion and a little thought shows that it is satisfied by $y = \sin x$. The exact solution is therefore known and the numerical solution can be compared with this. The final value of x is just larger than $\frac{1}{2}\pi$ and so the solution should be very close to 1 in this region. The first derivative is $\cos x$ and this should be almost zero when $x = 1.6$.

The numerical solution is calculated according to the method described above and so initially we have

$$x_0 = 0, y_0 = 0, y_0' = 1, \text{ and } y_0'' = -y_0 = 0$$

and subsequent values are calculated from the recurrence relationships

$$x_{r+1} = x_r + 0.1,$$

$$y_{r+1} = y_r + 0.1y_r' + 0.05y_r'',$$

$$y_{r+1}' = y_r' + 0.1y_r'',$$

$$y_{r+1}'' = -y_{r+1}.$$

A program in BASIC to perform the calculations and print x, y, y', and y'' is given below

```
10 PRINT "Y" + Y = 0"
20 PRINT
30 PRINT "X", "Y", "D1", "D2"
40 PRINT
50 X = 0
```

```
 60  Y = 0
 70  D1 = 1
 80  D2 = −Y
 90  PRINT X, Y, D1, D2
100  X = X + 0.1
110  IF X > 1.6 THEN 999
120  Y = Y + 0.1 * D1 + 0.005 * D2
130  D1 = D1 + 0.1 * D2
140  D2 = −Y
150  GOTO 90
999  END
```

In this program X represents x, Y represents y, and D1 and D2 represent the first and second derivatives y' and y''. The results printed out are

X	Y	D1	D2
0	0	1	−0
0.1	0.1	1	−0.1
0.2	0.1995	0.99	−0.1995
0.3	0.29750	0.97005	−0.29750
0.4	0.39302	0.94030	−0.39302
0.5	0.48508	0.90100	−0.48508
0.6	0.57276	0.85249	−0.57276
0.7	0.65514	0.79521	−0.65514
0.8	0.73139	0.72970	−0.73139
0.9	0.80070	0.65656	−0.80070
1.0	0.86236	0.57649	−0.86236
1.1	0.91569	0.49025	−0.91569
1.2	0.96014	0.39868	−0.96014
1.3	0.99521	0.30267	−0.99521
1.4	1.0205	0.20315	−1.0205
1.5	1.0357	0.10110	−1.0357
1.6	1.0406	−2.4708E−03	−1.0406

The final figure in the D1 column is −2.4708E–03 which means $−2.4708 \times 10^{-3}$ or −0.0024708.

It can be seen that the final value of Y is about 4 per cent too high. Greater accuracy can be obtained by reducing the step interval.

In order to ensure that the method is understood the first few rows of calculation should be done 'by hand'. It can also be of value to try and follow the program through line by line. This should be possible without any knowledge of BASIC.

Exercises

1. Use Euler's method to solve the differential equation

$$\frac{dy}{dx} = x, y = 0 \text{ when } x = 0,$$

for values of x from 0 to 1 in steps of 0.1. Solve the equation analytically and compare the numerical results with exact values at these points.

2. Solve the differential equation

$$\frac{dy}{dx} = x - y, y = 0 \text{ when } x = 0$$

by Euler's method for values of x from 0 to 1 in steps of 0.1.

3. Solve numerically

$$\frac{dy}{dx} = y, y = 1 \text{ when } x = 0$$

for values of x from 0 to 1 in steps of 0.1. Compare these results with the exact values and note that they are getting progressively worse. Can you find an explanation for this?

4. Solve numerically

$$\frac{dy}{dx} = x^2 + xy, y = 4 \text{ when } x = 2,$$

for values of x from 2 to 2.5 in steps of 0.05. Also solve the equation for values of x from 1.5 to 2 in steps of 0.1.

5. Solve numerically the second-order equation

$$\frac{d^2y}{dx^2} = x - y, y = -1, \frac{dy}{dx} = 1 \text{ when } x = 0$$

for values of x between 0 and 1 in steps of 0.1.

Appendix A

A.1. Real numbers

THE HEADING 'real numbers' seems to imply that there are other numbers which might be called 'unreal numbers'. Such a division of numbers does exist with a second type being called 'imaginary' (cf. Appendix A.3). The *real numbers* can broadly be regarded as the every-day numbers used in measurements and calculation.

The real numbers may be represented by points on a line often called the *real line*. An *origin* 0 is chosen and a *unit distance* defined. Such a line is shown in Figure A.1. The *sign* (+ or −) of a number determines which side of the origin it is with positive numbers being

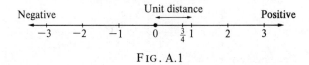

FIG. A.1

conventionally to the right. The *positive integers* (*whole numbers*) can be represented by points at multiples of the unit distance to the right of the origin. The *negative integers* lie at corresponding points to the left of the origin. Subdivisions of the unit distance can be made to give *fractions*. For example $\frac{3}{4}$ means divide the unit distance into four equal subunits and consider a point three such subunits to the right of the origin (Figure A.1).

The 'size' of a number is simply its distance from the origin irrespec-tive of direction left or right. In mathematics the 'size' is known as the *modulus* of the number. The terms absolute value and numerical value are also used for the modulus. The modulus is always positive unless the number itself happens to be zero in which case the modulus is zero. It is indicated by writing the number between two short vertical lines. Thus,

> the modulus of 7 is written |7| and is equal to 7;
> the modulus of −5 is written |−5| and is equal to 5;
> the modulus of 0 is written |0| and is equal to 0.

If a number a lies to the right of a number b on the real line then a is said to be greater than b. This is written $a > b$. The larger or open end of the symbol $>$ is towards the larger number. Hence $b < a$, meaning b is less than a, also means a is greater than b. Mathematical statements such as $a > b$ or $4x + 3y - 7 < 0$ are called *inequalities*. The symbols $\geqslant$, meaning greater than or equal to, and $\leqslant$, meaning less than or equal to, are also used. For example we may write, for any real number a,

$$|a| \geqslant a.$$

The condition of equality holds if a is positive or zero.

The *rational numbers* are defined to be the real numbers which can be expressed in the form m/n where m and n are positive or negative integers. The integer m may be zero but n cannot be zero. All decimals of a finite length and all recurring decimals are rational since they can be expressed in the form m/n, e.g. $0.4721 = 4721/10\,000$, $0.1111\ldots = 0.\dot{1} = \frac{1}{9}$. Numbers which cannot be expressed in this form are called *irrational*. Examples are the non-recurring decimals such as $\sqrt{2} = 1.41421\ldots$, $\sqrt{3} = 1.7320\ldots$, $\pi = 3.14159\ldots$.

A.2. Indices

A number a multiplied by itself m times is written a^m where m is called the *index, power,* or *exponent* of the *base* number a.

If a^m is multiplied by a^n, a is being multiplied by itself a total of $(m + n)$ times. Hence

$$a^m \times a^n = a^{(m+n)}.$$

If a^m is divided by a^n, the m factors of a^m are reduced in number by n, the number of factors of a^n. The remaining number of factors is therefore $(m - n)$. Hence

$$a^m \div a^n = a^{(m-n)}.$$

If $n = m$, a^m is divided by itself and the result is 1. The right-hand side gives a^0 and so

$$a^0 = 1.$$

If n is greater than m $(n > m)$, the index $(m - n)$ will be negative. However the left-hand side will have $(n - m)$ factors a in its denominator after complete cancellation of the numerator. Hence

$$a^{(m-n)} = a^{-(n-m)} = 1/a^{(n-m)}.$$

More generally,

$$a^{-m} = \frac{1}{a^m}.$$

A number with a negative index is interpreted to mean the reciprocal of the number with the corresponding positive index, e.g.

$$7^{-2} = \frac{1}{7^2} = \frac{1}{49}.$$

The interpretation of negative indices has been developed to be consistent with the multiplication and division rules for positive integer indices.

A consistent interpretation can be sought for fractional indices. Consider how $a^{\frac{1}{2}}$ might be interpreted. If it is assumed that the same rules which hold for positive integer indices are to hold for fractional indices then we may write

$$a^{\frac{1}{2}} \times a^{\frac{1}{2}} = a^{(\frac{1}{2} + \frac{1}{2})} = a^1 = a.$$

The number $a^{\frac{1}{2}}$ is the number which when multiplied by itself gives a. It is therefore the square root of a.

$$a^{\frac{1}{2}} = \sqrt{a}.$$

Similarly,

$$a^{\frac{1}{n}} = \sqrt[n]{a}.$$

If a^m is multiplied by itself n times the total number of times that a is multiplied by itself is $m \times n$. Hence

$$(a^m)^n = a^{mn}.$$

This same rule gives the result

$$(a^{1/n})^m = a^{m/n}$$

which enables an interpretation of the general fractional index m/n to be made.

$$a^{m/n} = (\sqrt[n]{a})^m = \sqrt[n]{a^m}.$$

It does not matter whether the root or the power is evaluated first.

Indices provide a compact way of writing very small or very large numbers. The way in which they are written is known as *scientific notation*. The significant figures are quoted first as a number between 1 and 10 followed by an appropriate power of 10 to increase or decrease the number to the correct magnitude. For example, the speed of light *in vacuo* is 299 800 000 m s^{-1}, which can be written 2.998 $\times$ 10^8 m s^{-1}. The charge on an electron is

$$0.000\ 000\ 000\ 000\ 000\ 000\ 160\ 2C$$

which in scientific notation becomes

$$1.602 \times 10^{-19} \text{ C.}$$

Avogadro's number is

$$602\ 500\ 000\ 000\ 000\ 000\ 000\ 000$$

and becomes

$$6.025 \times 10^{23}.$$

Apart from being concise scientific notation has the advantage of enabling the order of magnitude of any number to be readily seen and compared with others. It is also clear how many significant figures are being quoted.

A.3. Solution of the quadratic equation

The general quadratic equation in x takes the form

$$ax^2 + bx + c = 0$$

where a, b, and c are constant coefficients and $a \neq 0$. This equation can be solved for x by a method known as 'completing the square'. It involves making the terms containing x into a perfect square and then taking the square root. The resulting linear expression can easily be solved for x.

The equation can be written

$$x^2 + \frac{b}{a}x = -\frac{c}{a}.$$

Adding the square of half the coefficient of x to both sides makes the left-hand side into a perfect square.

$$x^2 + \frac{b}{a}x + \left(\frac{b}{2a}\right)^2 = \left(x + \frac{b}{2a}\right)^2 = -\frac{c}{a} + \left(\frac{b}{2a}\right)^2.$$

Therefore,

$$x + \frac{b}{2a} = \pm\sqrt{\left\{\left(\frac{b}{2a}\right)^2 - \frac{c}{a}\right\}}$$

$$= \pm\frac{\sqrt{(b^2 - 4ac)}}{2a}.$$

Therefore,

$$x = \frac{-b \pm \sqrt{(b^2 - 4ac)}}{2a}.$$

This is *the formula* so often quoted for the solution of a quadratic equation. The expression inside the square root, $b^2 - 4ac$, is known as the *discriminant*. If it is positive, its square root can be found and two distinct solutions of the quadratic equation exist. If it is zero, the square root vanishes and only a single solution occurs. If it is negative the square root cannot be taken and solutions do not exist in terms of conventional (real) numbers.

This last case in which the discriminant is negative can be overcome in a mathematically formal sense by the introduction of a number i. The only property required of this number in order to deal with a negative discriminant is that its square shall equal -1. Hence, i may be defined by the equation

$$i^2 = -1.$$

Using this definition the square roots of negative numbers can, in a formal sense, be found, e.g.

$$\sqrt{-4} = \sqrt{4i^2} = \pm 2i.$$

A number which contains a factor i is called an *imaginary* number to distinguish it from the real numbers of everyday life.

Example. Solve the quadratic equation $x^2 + 4x + 13 = 0$.

The formula gives

$$x = \frac{-4 \pm \sqrt{(4^2 - 4 \times 1 \times 13)}}{2} = -2 \pm \sqrt{-9}$$

$$= -2 \pm \sqrt{9i^2} = -2 \pm 3i.$$

This solution (or pair of solutions) contains a real number and an imaginary number. The solution is said to be *complex*. A *complex number* has the form $a + ib$ where a is a real number known as the *real part* and b is also a real number known as the *imaginary part*.

A.4. Series

A *series* is formed by adding together a sequence of terms each of which has been produced by applying a given rule to consecutive positive integers, perhaps including zero. One of the simplest series consists of the sum of the positive integers themselves,

$$1 + 2 + 3 + \ldots + i + \ldots + n.$$

This series has n terms, the ith or general terms being i itself. The general term is usually included since it indicates the rule from which the terms are calculated. Series need not be finite in length, an example being

$$1 + \frac{1}{4} + \frac{1}{9} + \frac{1}{16} + \ldots + \frac{1}{i^2} + \ldots.$$

This is an *infinite series* and such series may or may not add up to a finite sum. In this case it can be shown that the sum is $\frac{1}{6}\pi^2$. The summation of series, whether finite or infinite, is of considerable importance. If the series is finite and has n terms its sum is usually represented by S_n. If the series is infinite and can be summed its sum is represented by S_∞, the symbol ∞ meaning infinity.

Consider the series of the positive integers themselves.

$$S_n = 1 + 2 + 3 + 4 + \ldots + i + \ldots + n.$$

It may be written in a more compact form using the *sigma notation, Σ*. This symbol is a Greek capital letter sigma and is used to indicate a summation. The range of integers required and the general term are quoted in the following way.

$$S_n = \sum_{i=1}^{n} i.$$

This particular series can be summed by writing it down forwards and then backwards and adding the two together.

$$S_n = 1 + 2 + 3 + \ldots + (n-2) + (n-1) + n,$$

$$S_n = n + (n-1) + (n-2) + \ldots + 3 + 2 + 1.$$

Therefore,

$$2S_n = (n+1) + (n+1) + (n+1) + \ldots + (n+1) + (n+1) + (n+1).$$

There are n terms in this series and so

$$2S_n = n \times (n+1).$$

Therefore,
$$S_n = \frac{1}{2}n(n+1).$$

The sum of the squares of the first n integers can be shown to be given by

$$S_n = 1^2 + 2^2 + 3^2 + \ldots + i^2 + \ldots + n^2 = \sum_{i=1}^{n} i^2 = \frac{1}{6}n(n+1)(2n+1)$$

and the sum of their cubes by

$$S_n = 1^3 + 2^3 + 3^3 + \ldots + i^3 + \ldots + n^3 = \sum_{i=1}^{n} i^3 = \frac{1}{4}n^2(n+1)^2.$$

The proofs of these summations are somewhat more complicated and are not included here.

The *geometric progression* or *series* is given by

$$a + ar + ar^2 + \ldots + ar^i + \ldots + ar^{n-1}$$

where r is known as the common ratio. In this case the integers used run from $i = 0$ to $i = n - 1$ and so the series has n terms. Summation can be achieved in a relatively simple manner as follows.

$$S_n = a + ar + ar^2 + \ldots + ar^i + \ldots + ar^{n-1}$$
$$rS_n = \phantom{a + {}} ar + ar^2 + \ldots + ar^i + \ldots + ar^{n-1} + ar^n.$$

Subtraction then gives

$$(1 - r)S_n = a - ar^n$$

since all other terms cancel out. Hence

$$S_n = \sum_{i=0}^{n-1} ar^i = \frac{a(1 - r^n)}{1 - r} = \frac{a(r^n - 1)}{r - 1},$$

the two forms being used for $|r| < 1$ and $|r| > 1$ respectively, although they are identical. If $r = 1$, it can be seen from the series itself that

$$S_n = na.$$

If $r = -1$ then

$$S_n = 0 \quad \text{if } n \text{ is even},$$
$$= a \quad \text{if } n \text{ is odd}.$$

It should be noted that if $|r| < 1$, i.e. $-1 < r < +1$, the corresponding infinite series,

$$a + ar + ar^2 + \ldots + ar^i + \ldots.$$

can also be summed. To see this consider the formula for S_n above and note what happens as $n \to \infty$ (n tends to infinity) with $|r| < 1$. The term r^n in the formula becomes smaller and smaller as n increases and finally as n approaches infinity it vanishes. Hence

$$S_\infty = \frac{a}{1 - r}, \; |r| < 1.$$

If $r = \frac{1}{2}$, for example, and $a = 1$, then

$$S_\infty = 1 + \frac{1}{2} + \frac{1}{4} + \frac{1}{8} + \frac{1}{16} + \ldots = \frac{1}{1 - \frac{1}{2}} = 2.$$

A.5. Binomial theorems

Direct multiplication shows that

$$
\begin{aligned}
(a + b)^0 &= 1 \\
(a + b)^1 &= a + b \\
(a + b)^2 &= a^2 + 2ab + b^2 \\
(a + b)^3 &= a^3 + 3a^2b + 3ab^2 + b^3 \\
(a + b)^4 &= a^4 + 4a^3b + 6a^2b^2 + 4ab^3 + b^4 \\
(a + b)^5 &= a^5 + 5a^4b + 10a^3b^2 + 10a^2b^3 + 5ab^4 + b^5 \\
(a + b)^6 &= a^6 + 6a^5b + 15a^4b^2 + 20a^3b^3 + 15a^2b^4 + 6ab^5 + b^6.
\end{aligned}
$$

The coefficients involved in the expansion of $(a + b)^n$ form *Pascal's triangle.*

n	Coefficients
0	1
1	1 1
2	1 2 1
3	1 3 3 1
4	1 4 6 4 1
5	1 5 10 10 5 1
6	1 6 15 20 15 6 1

In each row the numbers can be formed by adding together the appropriate pair of numbers from the row above as indicated by the brackets. Each row starts and ends with a 1 and the next to first and next to last numbers are n. The numbers are known as the *binomial coefficients* and in the general case are denoted by ^nC_i (or $_nC_i$ or $\binom{n}{i}$) where

$$^nC_i = \frac{n!}{i!(n-i)!}.$$

$n! = n \times (n-1) \times (n-2) \times \ldots \times 3 \times 2 \times 1$ and $0! = 1$ by definition in order to maintain consistency. Hence $1! = 1$, $2! = 2 \times 1 = 2$, $3! = 3 \times 2 \times 1 = 6$, $4! = 4 \times 3 \times 2 \times 1 = 24$, etc.

The *binomial theorem* for a positive integer index n may be written

$$(a + b)^n = \sum_{i=0}^{n} {}^nC_i a^{n-i} b^i$$

$$= {}^nC_0 a^n + {}^nC_1 a^{n-1} b + {}^nC_2 a^{n-2} b^2 + \ldots + {}^nC_{n-1} ab^{n-1} + {}^nC_n b^n.$$

Using the expression for ${}^nC_i$,

$${}^nC_i = \frac{n!}{i!(n-i)!} = \frac{n!}{(n-i)!i!} = \frac{n!}{(n-i)!(n-(n-i))!} = {}^nC_{n-i}$$

which shows that the binomial coefficients are symmetrical, a feature clear from the direct expansions. In particular

$${}^nC_0 = \frac{n!}{0!(n-0)!} = \frac{n!}{n!} = 1 = {}^nC_n$$

and

$${}^nC_1 = \frac{n!}{1!(n-1)!} = \frac{n!}{(n-1)!} = n = {}^nC_{n-1}.$$

The formation of Pascal's triangle can be seen by considering two consecutive binomial coefficients ${}^nC_i$ and ${}^nC_{i+1}$ where the index is n.

$${}^nC_i + {}^nC_{i+1} = \frac{n!}{i!(n-i)!} + \frac{n!}{(i+1)!(n-i-1)!}$$

$$= \frac{(i+1)n! + (n-i)n!}{(i+1)!(n-i)!} = \frac{(n+1)n!}{(i+1)!(n+1-(i+1))!}$$

$$= {}^{n+1}C_{i+1}.$$

This is the corresponding coefficient with index $n + 1$.

As an application of the binomial theorem consider the expansion of $(2 + x)^7$.

$$(2 + x)^7 = \sum_{i=0}^{7} {}^7C_i(2)^{7-i}(x)^i$$

$$= {}^7C_0 2^7 + {}^7C_1 2^6 x + {}^7C_2 2^5 x^2 + {}^7C_3 2^4 x^3 + {}^7C_4 2^3 x^4$$

$$+ {}^7C_5 2^2 x^5 + {}^7C_6 2x^6 + {}^7C_7 x^7$$

$$= 2^7 + 7 \times 2^6 x + \frac{7!}{2!5!} 2^5 x^2 + \frac{7!}{3!4!} 2^4 x^3 + \frac{7!}{4!3!} 2^3 x^4$$

$$+ \frac{7!}{5!2!} 2^2 x^5 + 7 \times 2x^6 + x^7$$

$$= 2^7 + 7 \times 2^6 x + 21 \times 2^5 x^2 + 35 \times 2^4 x^3 + 35 \times 2^3 x^4$$

$$+ 21 \times 2^2 x^5 + 7 \times 2x^6 + x^7$$

$$= 128 + 448x + 672x^2 + 560x^3 + 280x^4 + 84x^5$$

$$+ 14x^6 + x^7.$$

For n a positive integer the binomial expansion consists of $(n + 1)$ terms.

If the index α of $(a + b)^\alpha$ is not a positive integer an expansion can still be performed but it is infinite and its validity depends upon the relative magnitudes of a and b. If it is assumed that $|a| > |b|$ then we may write

$$(a + b)^\alpha = a^\alpha \left(1 + \frac{b}{a}\right)^\alpha$$

where $|b/a| < 1$. This can be expanded using the *binomial theorem* for non-positive integer (negative and fractional) indices which takes the form

$$(1 + x)^\alpha = 1 + \alpha x + \frac{\alpha(\alpha - 1)}{2!} x^2 + \frac{\alpha(\alpha - 1)(\alpha - 2)}{3!} x^3 + \dots ,$$

where it is required that $|x| < 1$ if the expansion is to be valid.

The expansion of $\sqrt{(4 + x)}$ can be found, for example, by writing

$$\sqrt{(4+x)} = 2\left(1 + \frac{x}{4}\right)^{\frac{1}{2}} = 2\left\{1 + \frac{1}{2}\left(\frac{x}{4}\right) + \frac{(\frac{1}{2})(-\frac{1}{2})}{2!}\left(\frac{x}{4}\right)^2 + \ldots\right\}$$

$$= 2 + \frac{x}{4} - \frac{x^2}{64} + \ldots$$

From this an estimate of $\sqrt{3}$ can be made by putting $x = -1$.

$$\sqrt{3} \approx 2 - \frac{1}{4} - \frac{1}{64} = 1.734.$$

The inclusion of more terms would have produced greater accuracy.

A.6. Method of least squares

A situation often arises in which a set of n experimental results (x_i, y_i), $i = 1, 2, 3, \ldots, n$, has been produced and the relationship between x and y is believed to be linear. A plot of such a set of results might look like Figure A.2. A 'best straight line' is drawn and its slope and intercept determined in order to produce the equation $y = mx + c$.

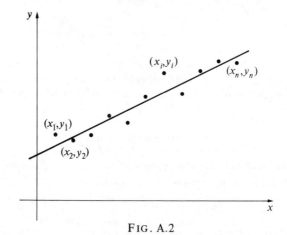

FIG. A.2

From this it may be hoped that some empirical law can be found or some hypothesis tested. The line is drawn in such a way that it passes 'as near as possible' to all the experimental points. This is a rather vague criterion to operate and something more precise would be desirable.

Consider an arbitrary experimental point (x_i, y_i). The independent variable x_i is chosen by the observer and is assumed to have no error. Any error which does occur will be in the observed experimental value, y_i, the dependent variable. The difference d_i between this value and the corresponding value on the line $y = mx + c$, given by $mx_i + c$, is a measure of how well the line fits the experimental point. The geometrical significance of d_i is shown in Figure A.3. It is the vertical

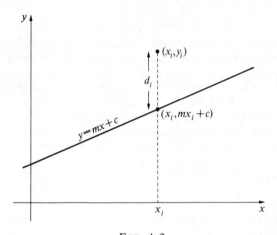

F IG. A.3

distance between the line and the experimental point. If the line passes through the point then $d_i = 0$. If all such points d_i, $i = 1, 2, \ldots, n$, are considered the line is a perfect fit if all $d_i = 0$. If it is not a perfect fit then some $d_i \neq 0$ and a function of the d_i can be used to judge how good the fit is. The sum of all the d_i is such a function which is certainly zero when the fit is perfect. However, if two points lie off the line by equal amounts but on opposite sides of the line the corresponding d_i will be equal in magnitude but opposite in sign. Their sum will be zero giving the false impression of a perfect fit. For this and other reasons the criterion of the *sum of the squares* of all the d_i is employed. This function of the d_i is given by

$$D = \sum_{i=1}^{n} d_i^2 = d_1^2 + d_2^2 + d_3^2 + \ldots + d_i^2 + \ldots + d_n^2$$

and can be used as a measure of how well any particular line, $y = mx + c$, fits the experimental data. Conversely, the two parameters, slope m and intercept c, can be varied in order to minimize D. This is known as the *method of least squares* and the resulting values of m and

c give the line of best fit according to the *least squares* criterion.

From what has been said previously and from Figure A.3,

$$d_i = y_i - (mx_i + c)$$

and so

$$D = \sum_{i=1}^{n} (y_i - mx_i - c)^2$$

which is a function of the two parameters m and c. We wish to find the values of m and c which correspond to a minimum value of D. In §4.6 maxima and minima of functions of a single variable are discussed and it is shown that at maxima and minima the slope dy/dx of the curve $y = f(x)$ is zero. Here we have a function D which depends upon both m and c. Methods akin to those employed in §4.6 do exist for such functions and involve equating the two *partial derivatives* $\partial D/\partial m$ and $\partial D/\partial c$ to zero. The partial derivative $\partial D/\partial m$ means that D is differentiated with respect to m (cf. Chapter 4) with any other variables being regarded as constants. Equating the two partial derivatives to zero leads to two equations from which m and c can be determined. Whether a maximum or minimum has been found requires some further investigation. Intuition tells us that a very badly fitting line can always be drawn which makes D large and so there appears to be no upper bound on D. It seems likely therefore that solving $\partial D/\partial m = 0$ and $\partial D/\partial c = 0$ will lead only to a minimum. This is indeed the case but it is not proved here.

Now

$$D = \sum_{i=1}^{n} (y_i - mx_i - c)^2$$

and so

$$\frac{\partial D}{\partial m} = \sum_{i=1}^{n} 2(y_i - mx_i - c)(-x_i)$$

and

$$\frac{\partial D}{\partial c} = \sum_{i=1}^{n} 2(y_i - mx_i - c)(-1).$$

The condition $\partial D / \partial c = 0$ gives

$$\sum_{i=1}^{n} y_i - m \sum_{i=1}^{n} x_i - c \sum_{i=1}^{n} 1 = 0$$

and so

$$c = \frac{1}{n} \sum_{i=1}^{n} y_i - \frac{m}{n} \sum_{i=1}^{n} x_i.$$

The condition $\partial D / \partial m = 0$ yields

$$\sum_{i=1}^{n} x_i y_i - m \sum_{i=1}^{n} x_i^2 - c \sum_{i=1}^{n} x_i = 0$$

and substituting in for c gives

$$\sum_{i=1}^{n} x_i y_i - m \sum_{i=1}^{n} x_i^2 - \left\{ \frac{1}{n} \sum_{i=1}^{n} y_i - \frac{m}{n} \sum_{i=1}^{n} x_i \right\} \sum_{i=1}^{n} x_i = 0.$$

Note that the mean values of the x_i and y_i are given respectively by

$$\bar{x} = \frac{1}{n} \sum_{i=1}^{n} x_i \quad \text{and} \quad \bar{y} = \frac{1}{n} \sum_{i=1}^{n} y_i.$$

Hence

$$\sum_{i=1}^{n} x_i y_i - m \sum_{i=1}^{n} x_i^2 - \{ \bar{y} - m\bar{x} \} \sum_{i=1}^{n} x_i = 0$$

and so

$$m = \frac{\displaystyle\sum_{i=1}^{n} x_i y_i - \bar{y} \sum_{i=1}^{n} x_i}{\displaystyle\sum_{i=1}^{n} x_i^2 - \bar{x} \sum_{i=1}^{n} x_i}.$$

$$= \frac{\dfrac{1}{n} \displaystyle\sum_{i=1}^{n} x_i y_i - \bar{x}\bar{y}}{\dfrac{1}{n} \displaystyle\sum_{i=1}^{n} x_i^2 - \bar{x}^2}.$$

Once m has been determined c is given by

$$c = \bar{y} - m\bar{x}.$$

The slope and intercept of the line which best fits the experimental points according to the least squares criterion can therefore be found analytically from the experimental points themselves.

The equation for c may be rewritten $\bar{y} = m\bar{x} + c$ showing that this line passes through the point $(\bar{x}, \bar{y})$ given by the means of the x_i and of the y_i respectively.

Appendix B

B.1. Basic formulae

Constants

$$\pi = 3.1415926535\ldots$$
$$e = 2.7182818284\ldots$$

Velocity of light in a vacuum	$c_0 = 2.997925 \times 10^8\,\mathrm{m\,s^{-1}}$
Permeability of a vacuum	$\mu_0 = 4\pi \times 10^{-7}\,\mathrm{kg\,m\,s^{-2}\,A^{-2}}$
Permittivity of a vacuum	$\epsilon_0 = 8.854185 \times 10^{-12}\,\mathrm{kg^{-1}\,m^{-3}\,s^4\,A^2}$
Rest mass of an electron	$m_e = 9.1091 \times 10^{-31}\,\mathrm{kg}$
Charge of an electron	$e = 1.60210 \times 10^{-19}\,\mathrm{C}$
Ice point	$T_{ice} = 273.150\,\mathrm{K}$
Gas constant	$R = 8.3143\,\mathrm{J\,K^{-1}\,mol^{-1}}$
Avagadro's number	$N_A = 6.02252 \times 10^{23}\,\mathrm{mol^{-1}}$
Molar volume of ideal gas at STP	$V_0 = 2.24136 \times 10^{-2}\,\mathrm{m^3\,mol^{-1}}$

Areas

Rectangle of height a and base b.

$$\text{Area} = ab.$$

Parallelogram of height h and base b.

$$\text{Area} = bh = ab \sin \phi.$$

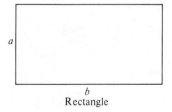

Rectangle

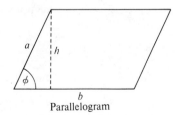

Parallelogram

Trapezium of height h and parallel sides a and b.

$$\text{Area} = \tfrac{1}{2}(a + b)h.$$

Triangle of height h and base b.

$$\text{Area} = \tfrac{1}{2}bh = \tfrac{1}{2}ab \sin \phi.$$

Circle of radius r.

$$\text{Area} = \pi r^2, \text{ circumference } 2\pi r.$$

Sector of a circle of radius r.

$$\text{Area} = \tfrac{1}{2}r^2 \phi, \ \phi \text{ in radians.}$$

Ellipse of semi-major axis a and semi-minor axis b.

$$\text{Area} = \pi ab.$$

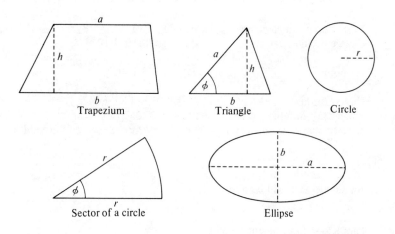

Trapezium   Triangle   Circle

Sector of a circle   Ellipse

Volumes

Rectangular parallelepiped with sides a, b, and c.

$$\text{Volume} = abc.$$

Sphere of radius r.

$$\text{Volume} = \tfrac{4}{3}\pi r^3, \text{ Surface area} = 4\pi r^2.$$

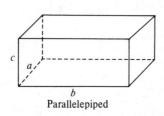

Parallelepiped

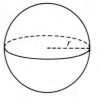

Sphere

Right circular cylinder of radius r and height h.

$$\text{Volume} = \pi r^2 h.$$

Right circular cone of radius r and height h.

$$\text{Volume} = \tfrac{1}{3}\pi r^2 h.$$

Pyramid of base area A and height h.

$$\text{Volume} = \tfrac{1}{3}A h.$$

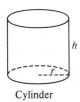

Cylinder

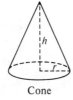

Cone

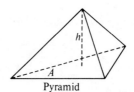

Pyramid

Indices (§A.2)

m factors $a \times a \times a \times \ldots \times a = a^m$

$a^m \times a^n = a^{m+n}$, $a^m \div a^n = a^{m-n}$, $a^0 = 1$,

$(a^m)^n = a^{mn}$, $a^{-m} = \dfrac{1}{a^m}$, $a^{1/n} = \sqrt[n]{a}$,

$a^{m/n} = \sqrt[n]{a^m} = (\sqrt[n]{a})^m$, $(ab)^m = a^m b^m$, $\left(\dfrac{a}{b}\right)^m = \dfrac{a^m}{b^m}$.

Logarithms (§2.7)

If $a^y = x$ then $y = \log_a x$, $a > 0, a \neq 1$.

$\log_a 1 = 0$, $\log_a a = 1$

$$\log_a x + \log_a y = \log_a (xy), \quad \log_a x - \log_a y = \log_a\left(\frac{x}{y}\right),$$

$$\log_a \frac{1}{x} = -\log_a x, \quad \log_a x^n = n \log_a x,$$

$$\log_b a = \frac{1}{\log_a b}, \quad \log_b x = \frac{\log_a x}{\log_a b}.$$

$$\log_a(a^x) = x, \quad a^{\log_a x} = x.$$

Trigonometric functions (§2.8)

$$\sin^2 \phi + \cos^2 \phi = 1, \quad 1 + \tan^2 \phi = \sec^2 \phi,$$

$$\sin(\phi \pm \theta) = \sin \phi \cos \theta \pm \cos \phi \sin \theta,$$
$$\cos(\phi \pm \theta) = \cos \phi \cos \theta \mp \sin \phi \sin \theta,$$

$$\tan(\phi \pm \theta) = \frac{\tan \phi \pm \tan \theta}{1 \mp \tan \phi \tan \theta},$$

$$\sin 2\phi = 2 \sin \phi \cos \phi, \quad \cos 2\phi = \cos^2 \phi - \sin^2 \phi,$$

$$\tan 2\phi = \frac{2 \tan \phi}{1 - \tan^2 \phi}$$

$$\sin 3\phi = 3 \sin \phi - 4 \sin^3 \phi, \quad \cos 3\phi = 4 \cos^3 \phi - 3 \cos \phi,$$

$$\tan 3\phi = \frac{3 \tan \phi - \tan^3 \phi}{1 - 3 \tan^2 \phi}$$

$$\sin \phi = \frac{2 \tan (\phi/2)}{1 + \tan^2 (\phi/2)}, \quad \cos \phi = \frac{1 - \tan^2 (\phi/2)}{1 + \tan^2 (\phi/2)}.$$

Series

Arithmetic series

$$a + (a + d) + (a + 2d) + \ldots + (a + id) + \ldots + (a + (n - 1)d)$$

$$= \tfrac{1}{2}n(2a + (n - 1)d).$$

Geometric series

$$a + ar + ar^2 + \ldots + ar^i + \ldots + ar^{n-1} = \frac{a(1 - r^n)}{1 - r}.$$

$$a + ar + ar^2 + \ldots + ar^i + \ldots = \frac{a}{1 - r}, \ |r| < 1.$$

Sums of powers of positive integers

$$1 + 2 + 3 + \ldots + i + \ldots + n = \tfrac{1}{2}n(n + 1),$$

$$1^2 + 2^2 + 3^2 + \ldots + i^2 + \ldots + n^2 = \tfrac{1}{6}n(n + 1)(2n + 1),$$

$$1^3 + 2^3 + 3^3 + \ldots + i^3 + \ldots + n^3 = \tfrac{1}{4}n^2(n + 1)^2,$$

$$1^4 + 2^4 + 3^4 + \ldots + i^4 + \ldots + n^4 = \tfrac{1}{3}n(n + 1)(2n + 1)(3n^2 + 3n - 1).$$

Series expansions

Taylor series

$$f(x) = f(a) + (x - a)f'(a) + \frac{(x - a)^2}{2!}f''(a) + \ldots$$

Maclaurin series

$$f(x) = f(0) + xf'(0) + \frac{x^2}{2!}f''(0) + \ldots$$

Binomial series

$$(a + b)^n = \sum_{i=0}^{n} {}^nC_i a^{n-i}b^i, \quad {}^nC_i = \frac{n!}{i!(n - i)!}$$

$$(1 + x)^\alpha = 1 + \alpha x + \frac{\alpha(\alpha - 1)}{2!}x^2 + \frac{\alpha(\alpha - 1)(\alpha - 2)}{3!}x^3 + \ldots$$

where α is not a positive integer and $|x| < 1$.

Some examples of series expansions are:

$$e^x = 1 + x + \frac{x^2}{2!} + \frac{x^3}{3!} + \ldots,$$

$$a^x = e^{x \ln a} = 1 + x \ln a + \frac{(x \ln a)^2}{2!} + \frac{(x \ln a)^3}{3!} + \ldots,$$

$$\ln(1 + x) = x - \frac{x^2}{2} + \frac{x^3}{3} - \frac{x^4}{4} + \ldots, \quad -1 < x \leqslant 1,$$

$$\sin x = x - \frac{x^3}{3!} + \frac{x^5}{5!} - \frac{x^7}{7!} + \ldots,$$

$$\cos x = 1 - \frac{x^2}{2!} + \frac{x^4}{4!} - \frac{x^6}{6!} + \ldots,$$

$$\tan x = x + \frac{x^3}{3} + \frac{2x^5}{15} + \frac{17x^7}{315} + \ldots, \quad |x| < \frac{\pi}{2}.$$

B.2. Derivatives

In the following a and b are constants and u and v are functions of x.

$$y = a \qquad\qquad \frac{dy}{dx} = 0$$

$$y = x^\alpha \qquad\qquad \frac{dy}{dx} = \alpha x^{\alpha - 1}$$

$$y = (ax + b)^\alpha \qquad\qquad \frac{dy}{dx} = \alpha a (ax + b)^{\alpha - 1}$$

$$y = e^x \qquad\qquad \frac{dy}{dx} = e^x$$

$$y = e^{ax} \qquad\qquad \frac{dy}{dx} = a e^{ax}$$

$$y = a^x \qquad\qquad \frac{dy}{dx} = a^x \ln a$$

$y = \ln x$ $\qquad\qquad \dfrac{dy}{dx} = \dfrac{1}{x}$

$y = \ln(ax + b)$ $\qquad\qquad \dfrac{dy}{dx} = \dfrac{a}{ax + b}$

$y = \log_a x$ $\qquad\qquad \dfrac{dy}{dx} = \dfrac{1}{x} \log_a e$

$y = \sin x$ $\qquad\qquad \dfrac{dy}{dx} = \cos x$

$y = \sin(ax + b)$ $\qquad\qquad \dfrac{dy}{dx} = a \cos(ax + b)$

$y = \cos x$ $\qquad\qquad \dfrac{dy}{dx} = -\sin x$

$y = \cos(ax + b)$ $\qquad\qquad \dfrac{dy}{dx} = -a \sin(ax + b)$

$y = \tan x$ $\qquad\qquad \dfrac{dy}{dx} = \sec^2 x$

$y = \tan(ax + b)$ $\qquad\qquad \dfrac{dy}{dx} = a \sec^2(ax + b)$

$y = \sec x$ $\qquad\qquad \dfrac{dy}{dx} = \sec x \tan x$

$y = \operatorname{cosec} x$ $\qquad\qquad \dfrac{dy}{dx} = -\operatorname{cosec} x \cot x$

$y = \cot x$ $\qquad\qquad \dfrac{dy}{dx} = -\operatorname{cosec}^2 x$

$y = \sin^{-1} x$ $\qquad\qquad \dfrac{dy}{dx} = \dfrac{1}{\sqrt{(1 - x^2)}}$

$y = \tan^{-1} x$ $\qquad\qquad \dfrac{dy}{dx} = \dfrac{1}{1 + x^2}$

$$y = u \pm v \qquad \frac{dy}{dx} = \frac{du}{dx} \pm \frac{dv}{dx}$$

$$y = uv \qquad \frac{dy}{dx} = u\frac{dv}{dx} + v\frac{du}{dx}$$

$$y = \frac{u}{v} \qquad \frac{dy}{dx} = \frac{v\dfrac{du}{dx} - u\dfrac{dv}{dx}}{v^2}$$

$$y = f(u) \qquad \frac{dy}{dx} = \frac{df}{du} \cdot \frac{du}{dx}$$

B.3. Integrals

In the following a and b are constants and u and v are functions of x. C is a constant of integration.

$$\int x^\alpha dx = \frac{x^{\alpha+1}}{\alpha+1} + C, \alpha \neq -1$$

$$\int \frac{dx}{x} = \ln x + C$$

$$\int (ax + b)^\alpha dx = \frac{(ax + b)^{\alpha+1}}{a(\alpha + 1)} + C, \alpha \neq -1$$

$$\int \frac{dx}{ax + b} = \frac{1}{a}\ln(ax + b) + C$$

$$\int e^x \, dx = e^x + C$$

$$\int e^{ax} dx = \frac{1}{a}e^{ax} + C$$

$$\int a^x dx = \frac{a^x}{\ln a} + C$$

$$\int \ln x \, dx = x \ln x - x + C$$

$$\int \log_a x \, dx = \log_a e(x \ln x - x) + C$$

$$\int \sin x \, dx = -\cos x + C$$

$$\int \sin(ax + b) \, dx = -\frac{1}{a} \cos(ax + b) + C$$

$$\int \cos x \, dx = \sin x + C$$

$$\int \cos(ax + b) \, dx = \frac{1}{a} \sin(ax + b) + C$$

$$\int \tan x \, dx = \ln(\sec x) + C = -\ln(\cos x) + C$$

$$\int \sec x \, dx = \ln(\sec x + \tan x) + C = \ln\left\{\tan\left(\frac{\pi}{4} + \frac{x}{2}\right)\right\} + C$$

$$\int \operatorname{cosec} x \, dx = \ln(\operatorname{cosec} x - \cot x) + C = \ln\left(\tan\frac{x}{2}\right) + C$$

$$\int \cot x \, dx = \ln(\sin x) + C$$

$$\int \frac{dx}{x^2 - a^2} = \frac{1}{2a} \ln\left(\frac{x - a}{x + a}\right) + C$$

$$\int \frac{dx}{a^2 - x^2} = \frac{1}{2a} \ln\left(\frac{a + x}{a - x}\right) + C$$

$$\int \frac{dx}{\sqrt{(a^2 - x^2)}} = \sin^{-1}\left(\frac{x}{a}\right) + C$$

$$\int \frac{dx}{a^2 + x^2} = \frac{1}{a} \tan^{-1}\left(\frac{x}{a}\right) + C$$

$$\int \frac{dx}{\sqrt{(x^2 + a^2)}} = \ln(x + \sqrt{(x^2 + a^2)}) + C$$

$$\int \frac{dx}{\sqrt{(x^2 - a^2)}} = \ln(x + \sqrt{(x^2 - a^2)}) + C$$

$$\int (u + v)\mathrm{d}x = \int u\mathrm{d}x \pm \int v\mathrm{d}x$$

$$\int u\mathrm{d}v = uv - \int v\mathrm{d}u$$

$$\int \frac{f'(x)}{f(x)}\,\mathrm{d}x = \ln(f(x)) + C$$

Answers to exercises

Exercises 1.5

1. (a) 3.56,   (b) 5.89×10^{-10}.
2. 16.45 million.
3. $k = \dfrac{1}{T}\ln 2$.
4. $y = a\sqrt{x} = ax^{\frac{1}{2}}$.
5. $y = ax^2, y = ax - b, y = a - bx, y = \dfrac{a}{x} = ax^{-1}$.

Exercises 2.2

2. (a) $y = x$,   (b) $y = \frac{1}{2}x + 2$ (c) $y = -2x$,

    (d) $y = -\frac{1}{3}x + \frac{11}{3}$,   (e) $y = 2$.

3. (a) $y = x + 1$,   (b) $y = 2x$,   (c) $y = -x + 4$,

    (d) $y = -\dfrac{b}{a}x + b$,   (e) $y = -x + (a + b)$,   (f) $y = 4$,

    (g) $y = 5$ (h) $x = 3$.

5. $y = -x + 4$, $(1, 3)$, $2\sqrt{2}$.

6. $\dfrac{4}{\sqrt{5}}$.

Exercises 2.3

2. (a) $y = x^2 - 2x - 3$, (b) $y = x^2 + 2x + 2$, (c) $y = 3 - x - 2x^2$,

    (d) $y = x^2 - 2$,   (e) $y = 4.5 + 2.4x - 3.2x^2$.

3. -1 and 3.
4. -5 and 2.
5. 56 weeks, 35 weeks before recordings began.
6. $200\ \text{m} \times 200\ \text{m} = 40\ 000\ \text{m}^2$, $200\ \text{m} \times 400\ \text{m} = 80\ 000\ \text{m}^2$.

Exercises 2.4

2. (a) $y = x^3 - 2x^2 + 3x - 4$,   (b) $y = x^3 - x$,

(c) $y = 10x^3 - 2x^2 + 1.5x - 2.7$.

3. $\dfrac{2}{3} \times \dfrac{2}{3} \times \dfrac{1}{6} = \dfrac{2}{27}\,\text{m}^3$.

4. $51\frac{1}{2}$ weeks.

5. $G = \dfrac{5}{81}T^3 - \dfrac{5}{3}T^2 + \dfrac{190}{9}T - 80$.

Exercises 2.5

2. Area $= x^2 + \dfrac{16}{x}$, $12\ \text{m}^2$ when $x = 2\ \text{m}$.

4. When $c = k_M, r = \frac{1}{2}R$.

Exercises 2.6

1. $m = 0.5 \times 2^{2t}$, $1.4074 \times 10^{14}\,\text{g}$.
2. 1.845×10^{11} tonnes, 3.689×10^{11} tonnes.
3. The exponential 2^x.
4. 0.693.
5. 0.693.
6. Carbon 14   1 g, 0.999966 g, 0.996274 g, 0.688439 g
   Cobalt 60   0.999641 g, 0.964804 g 0.019772 g, 0 g.
   Phosphorus 32   0.952684 g, 0.007850 g, 0 g, 0 g.
   Sodium 24   0.329877 g, 0 g, 0 g, 0 g.
7. 13.99 mins.
8. 2, 2.25, 2.48832, 2.593742, 2.704814, 2.718146, 2.718282, e.

Exercises 2.7

3. (a) 16,   (b) 114.49,   (c) 2.96249×10^9,
   (d) 4.66096,   (e) 0.0263041,   (f) 1.91885,
   (g) 0.518652,   (h) 0.0105652,   (i) 3.97099,
   (j) 0.830174.

4. (a) 0.993252,   (b) 1.85348,   (c) 2.74855,
   (d) $\overline{6}.94541$,   (e) 10.81235.

8. (a) 1.5850,   (b) 0.63091,   (c) 6,
   (d) 1.6826,   (e) 0.58495.

9. $\ln x = \dfrac{\log x}{0.43429}$.

10. (a) $x = \log y$,   (b) $x = \ln y$,   (c) $x = \frac{1}{5}\ln\!\left(\dfrac{y}{4}\right)$,

(d) $x = \ln(y - 2) - 3,$ (e) $x = \ln \{y \pm \sqrt{(y^2 - 1)}\}.$

11. $0.6065 : 1.$

Exercises 2.8

1. (a) $0.785,$ (b) $0.524,$ (c) $2.356,$
   (d) $0.698,$ (e) $1.396,$ (f) $1.745,$
   (g) $3.491,$ (h) $5.236,$ (i) $-1.571,$
   (j) $-0.175.$

2. (a) $60,$ (b) $45,$ (c) $5.730,$
   (d) $14.324,$ (e) $28.648,$ (f) $85.944,$
   (g) $114.592,$ (h) $343.775,$ (i) $-17.189,$
   (j) $-114.592.$

4. Approximate formulae for London are, Sunrise $= 5.9 + 2.2 \cos\left(\dfrac{n\pi}{26}\right),$

   Sunset $= 18.1 - 2.2 \cos\left(\dfrac{n\pi}{26}\right),$ Daylight $= 12.2 - 4.4 \cos\left(\dfrac{n\pi}{26}\right).$

5. Maximum of $\dfrac{\pi}{12}$ rad at $t = 0.614$ s and every 1.228 s thereafter.

   Period is 2.457 s.

6. $P_H = 60\,000 + 50\,000 \sin\left\{\dfrac{2\pi}{11}(t + 2)\right\},$ 2.98 years.

Exercises 2.9

1. (a) $-3,$ (b) $0,$ (c) $0,$ (d) $-4,$ (e) $5.$

Exercises 3.2

1. $x = 0,$ $y = \frac{1}{2},$ $z = -\frac{1}{2}.$
2. 49 cm.
3. $y = 2 - 3x.$
4. $A = 423M^{0.63} l^{0.11},$ 1555 cm$^2.$

Exercises 3.4

1. $I = 2.5 V^{0.75}.$
2. $M = 0.140B^{1.052}.$
3. $R = 512M^{0.724}.$

Exercises 3.5

1. $m = 0.400e^{0.03t}$, 1.792 g.
2. $\alpha = 16.7, I/I_0 = 0.535$.
3. $m = 175e^{-0.193t}$

Exercises 3.6

1. (a) $2x - 7$,   (b) $2x^2 - 5x - 3$,   (c) $3x^2 + 5x + 2$,
   (d) $x^3 - 3x^2$,   (e) $x^3 - 7x^2 + 14x - 8$.
2. $y = 2.409 - 0.950x - 0.378x^2$.
3. When $x = 2.60$, y should read 2.174.

Exercises 3.7

1. 2.14773,   2.14775.
2. $x = 0.130$, $y = 0.088, 0.099, 0.112$.
   $x = 0.170$, $y = 0.116, 0.122, 0.128$.
   $x = 0.280$, $y = 0.152, 0.155, 0.156$.
   $x = 0.550$, $y = 0.186, 0.187, 0.188$.
3. $t = 50$ min, $N = 2280, 2208$.
   $t = 57$ min, $N = 2812, 2760$.
   $N = 4000$, $t = 68$ min.
   $N = 7500$, $t = 87$ min.

Exercises 4.5

1. (a) 1,   (b) $2x$,   (c) $2(x + 1)$,

   (d) $4x + 1$,   (e) $2x - 4$,   (f) $\dfrac{-1}{(x + 1)^2}$,

   (g) $\dfrac{-2}{x^3}$,   (h) $\dfrac{1}{2\sqrt{(x + 4)}}$,   (i) $\dfrac{x + 1}{\sqrt{(x^2 + 2x - 3)}}$,

   (j) $\dfrac{-1}{2x\sqrt{x}}$.

2. (a) 3,   (b) $2x$,   (c) $3x^2 - 8x + 7$,

   (d) $2x + 1$,   (e) $5x^4 + 21x^2 - 10x$,
   (f) $7x^6 - 18x^5 + 35x^4 - 8x^3 + 18x^2 - 26x - 3$.
   (g) $\dfrac{x(x + 6)}{(x + 3)^2}$,   (h) $\dfrac{7}{(x + 4)^2}$,   (i) $\dfrac{5x - 7}{(x + 1)^3}$,

(j) $\dfrac{-x^6 + 14x^5 - 36x^4 + 20x^3 + 48x^2 - 100x + 55}{(x^4 - 3x^2 + 5)^2}$.

3. (a) $10(4x - 3)(2x^2 - 3x + 7)^9$,

(b) $\dfrac{3x^2 - 2}{2\sqrt{(x^3 - 2x + 3)}}$,

(c) $\frac{4}{3}(4x + 3)^{-\frac{2}{3}}$,

(d) $\frac{4}{5}(2x - 5)(x^2 - 5x + 7)^{-\frac{1}{5}}$,

(e) $-\frac{2}{3}(3x^2 - 2)(x^3 - 2x + 1)^{-\frac{5}{3}}$.

4. (a) $\dfrac{3x}{2y}$,

(b) $-\left(\dfrac{y}{x}\right)^{\frac{1}{3}}$,

(c) $\dfrac{2x - 3y}{3x - 4y}$,

(d) $\dfrac{y(3x - 2y)}{x(3y - 2x)}$,

(e) $-\dfrac{x}{y}$.

5. (a) $2e^{2x}$,

(b) $2e^{2x+1}$,

(c) $2(x + 1)e^{x^2+2x+1}$,

(d) $2^{2x+1}\ln 2$,

(e) $2x10^{x^2-3}\ln 10$,

(f) $\dfrac{1}{x + 1}$,

(g) $\dfrac{2(x + 1)}{x^2 + 2x - 1}$,

(h) $\cos x - \sin x$,

(i) $2\cos(2x)$,

(j) $3\cos x \sin^2 x$,

(k) $2x\cos(x^2)$,

(l) $x\cos x + \sin x$,

(m) $\dfrac{x\cos x - \sin x}{x^2}$,

(n) $-\tan x$,

(o) $-\csc x \cot x$,

(p) $\ln x$,   (q) $-(2x + 3)\cos(1 - 3x - x^2)$,

(r) $\dfrac{2\cos(4x + 1)}{\sqrt{\sin(4x + 1)}}$,

(s) $\dfrac{2(\cos 2x - x)}{\sin 2x - x^2}$,

(t) $\dfrac{3\ln 2}{3x - 1}2^{\ln(3x-1)}$.

Exercises 4.6

1. (a) $(0, 0)$ minimum, (b) $(-1\frac{1}{2}, 3\frac{1}{4})$ maximum,

(c) neither,

(d) $\left(-\dfrac{1}{\sqrt{3}}, \dfrac{2}{3\sqrt{3}}\right)$ max, $\left(\dfrac{1}{\sqrt{3}}, -\dfrac{2}{3\sqrt{3}}\right)$ min,

(e) neither,   (f) $(0, 0)$ max, $\left(-\dfrac{1}{\sqrt{2}}, -\dfrac{1}{4}\right)$ min, $\left(\dfrac{1}{\sqrt{2}}, -\dfrac{1}{4}\right)$ min,

(g) $(5, 75)$ min,   (h) $((2n + \frac{1}{2})\pi, 1)$max, $((2n + \frac{3}{2})\pi, -1)$min,

(i) (0.6435, 5) max together with other solutions,

(j) (0.6065, −0.1839)min.

2. $h = 2r$.

3. 2.

4. $\sqrt{3}$ km, 1.45 km from A (use a graph or Newton–Raphson §4.10).

6. 19.97 days, 52.5 millions.

Exercises 4.7

1. (a) $2x - \dfrac{4}{3}x^3 + \dfrac{4}{15}x^5$, (b) $1 + \dfrac{1}{3}x - \dfrac{1}{9}x^2$,

   (c) $1 - x^2 + \dfrac{x^4}{3}$, (d) $x + \dfrac{x^3}{3} + \dfrac{2}{15}x^5$,

   (e) $x - \dfrac{x^2}{2} + \dfrac{x^3}{3}$.

3. $\log(10 + x) \approx 1 + \dfrac{1}{\ln 10}\left(\dfrac{x}{10} - \dfrac{x^2}{200}\right)$.

4. $P \approx 10 + \dfrac{4}{5}t + \dfrac{3}{125}t^2$, when $t = 20$ error is 9.7%.

Exercises 4.8

1. 30 cm^3.

2. 1.25%.

3. 10.0066, 9.9667.

4. 3.5%.

5. $\dfrac{\mathrm{d}P}{\mathrm{d}t} = \tfrac{1}{5}\pi\sqrt{(-2 + 3P - P^2)}$, 77 000.

Exercises 4.9

1. −0.08 m^{3}h^{-1}.

2. −0.0048 cm s^{-1}.

3. 134, 94, −5, −86 Wm^{-2}h^{-1}.

Exercises 4.10

1. (a) 0 and 0.87672, (b) 0.51775, (c) 0.65292.

2. 5.74 years.

3. 0 and 9900.

Exercises 5.3

1. (a) $\dfrac{83}{6}$, (b) 42, (c) $4\frac{1}{2}$.

   (d) $e^2 - \dfrac{1}{e} = 7.0212$, (e) $1 + \dfrac{1}{\sqrt{2}}$.

2. (a) $\frac{1}{3}x^3 - 2x^2 + 3x + c$, (b) $2x^2 + \cos x + c$,

   (c) $\frac{2}{3}x^{\frac{3}{2}} + c$, (d) $x \sin x + \cos x + c$,

   (e) $\frac{1}{2}x^2 + \ln x + c$, (f) $(x^2 - 2x + 2)e^x + c$.

3. Area from $x = 0$ to $x = 2$ is $6\frac{2}{3}$, from $x = 2$ to $x = 4$ it is $-1\frac{1}{3}$.

4. $\frac{1}{3}$.

5. 40 000.

Exercises 5.5

1. (a) $\sqrt{(3 + 2x)} + c$, (b) $\frac{1}{6}(4x - 1)^{\frac{3}{2}} + c$, (c) $\dfrac{-1}{3(3x - 2)} + c$,

   (d) $\frac{1}{3}(x^2 - 1)^{\frac{3}{2}} + c$, (e) $-\frac{1}{2}\cos 2x + c$, (f) $\frac{1}{3}\sin^3 x + c$,

   (g) $\frac{2}{3}(\sin x)^{\frac{3}{2}} + c$, (h) $\frac{1}{2}\ln x + c$, (i) $\frac{1}{2}\ln(2x + 7) + c$,

   (j) $\frac{1}{2}\ln(x^2 + 2x - 5) + c$, (k) $\ln(\sec x) + c$,

   (l) $\ln(\sec x) + c$, (m) $\frac{1}{2}x\sqrt{(9 - 4x^2)} + \frac{9}{4}\sin^{-1}\left(\dfrac{2x}{3}\right) + c$,

   (n) $\frac{1}{3}\sin^{-1}\left(\dfrac{3x}{5}\right) + c$, (o) $\dfrac{1}{\sqrt{3}}\tan^{-1}(\sqrt{3}x)$.

2. (a) $(x - 1)e^x + c$, (b) $x \sin x + \cos x + c$,

   (c) $\frac{1}{4}x^2(2\ln x - 1) + c$, (d) $-x^2\cos x + 2x \sin x + 2\cos x + c$,

   (e) $\frac{1}{2}(\sin x - \cos x)e^x + c$, (f) $\dfrac{1}{140}(10x + 3)(4x - 3)^{\frac{5}{2}} + c$.

3. (a) $\ln\left(\dfrac{x - 2}{x - 1}\right) + c$, (b) $\frac{1}{2}x^2 + 5x + 18\ln(x - 3) + c$,

   (c) $x - 3\ln(x - 2) + 7\ln(x - 3) + c$,

(d) $\ln\left(\dfrac{x-1}{x-2}\right) - \dfrac{1}{x-2} + c,$

(e) $-\frac{1}{2}\ln(x-1) + \frac{1}{4}\ln(x^2+1) + \frac{5}{2}\tan^{-1}x + c,$

(f) $86\ln\left\{\dfrac{(x+2)^2}{x^2-3}\right\} - \dfrac{46}{x+2} - \dfrac{13}{2(x+2)^2} + \dfrac{149}{\sqrt{3}}\ln\left\{\dfrac{x-\sqrt{3}}{x+\sqrt{3}}\right\} + c.$

4. (a) 2,   (b) $\frac{2}{3}$,   (c) $1 - \dfrac{1}{\sqrt{2}}$,   (d) $1 - \dfrac{1}{e}$,   (e) 1.

5. Exponential: 17.655, 167.51, 9.9007×10^{11}.
   Quadratic: 18.047, 134.06, 8951.2.

6. 19.8 MJ.

Exercises 5.6

1.

	(a)	(b)	(c)	(d)	(e)
Trapezoidal	1.3933	1.9835	0.20689	0.80631	0.13213
Simpson	1.3865	2.0001	0.20686	0.80474	0.13216
Exact (5 fig)	1.3863	2.0000	0.20686	0.80472	0.13216

2.

Value of x	0	1	2	3	4	5
Simpson	0.42126	0.84135	0.96241	0.99865	0.99992	1.00000

3. 10.469 MJ.

Exercises 6.1

2. Yes, yes. $k = \frac{1}{2}\ln 2$.

3. Yes. $k = \dfrac{\ln 2}{5570}$.

Exercises 6.2

1.

	(a)	(b)	(c)	(d)	(e)	(f)	(g)	(h)	(i)	(j)	(k)	(l)
Order	1	2	2	1	2	2	3	1	1	1	1	2
Degree	1	1	1	1	1	1	1	1	1	2	1	1
Linear	Yes	Yes	Yes	Yes	No	No	Yes	Yes	Yes	No	No	Yes

Exercises 6.3

1. (a) $y = \frac{1}{2}x^2 + c$,   (b) $y = e^x$,   (c) $\frac{1}{3}(y+1)^3 = \frac{1}{4}x^4 + c$,

    (d) $e^y = e^t + c$,   (e) $\frac{1}{2}(y-1)^2 = x + \frac{1}{x} + 2$.

2. $P = 2(\frac{3}{2})^{t/2}$, 8.27 million.

3. $t = 5570 \dfrac{\log(0.01x)}{\log(0.5)}$ years.

4. 52.7 min, 45.2°C.

5. $P = \dfrac{90}{1 + 2(4)^{-\frac{t}{10}}}$, 60 million.

6. $C = \left(C_1 \ln \dfrac{r}{r_2} + C_2 \ln \dfrac{r_1}{r} \right) / \ln \dfrac{r_1}{r_2}$.

Exercises 6.4

1. (a) $y = A e^{-x} + x - 1$,   (b) $y = \frac{1}{16}(A e^{4x} + 12x + 3)$,

    (c) $y = \tan(x + c) - x$,   (d) $y = x - \dfrac{1 - e^{-2x}}{1 + e^{-2x}}$,

    (e) $y + \frac{1}{2}\sin\{2(x+y)\} - x = 0$.

2. $x^2 + y^2 = A e^{2x}$.

3. Put $ax + by + c = v$. $\dfrac{dv}{dx} = a + bf(v)$.

4. (a) $y = x(\ln x + c)$,   (b) $\tan^{-1}\left(\dfrac{y}{x}\right) = \frac{1}{2}\ln(x^2 + y^2) + c$,

    (c) $y^2 = 2x^2(\ln x + c)$,   (d) $x^2(x^2 - 2y^2) = 1$,

    (e) $y = x \sin^{-1}(x)$.

Exercises 6.5

1. (a) $y = c e^{-ax}$,   (b) $y = \frac{3}{2} + c e^{-2x}$,   (c) $y = \frac{3}{2} + c e^{-x^2}$,

    (d) $y = x + \dfrac{c}{x^2}$,   (e) $y = \dfrac{x-1}{x+1}e^x + \dfrac{c}{x^2 - 1}$,

    (f) $y = e^x(x^2 - 2x + 2)\cos x + c \cos x$.

4. $\dfrac{1}{y^3} = \frac{4}{3}e^{3x} - x - \frac{1}{3}$.

5. $T = 20 + 80(\frac{5}{8})^{\frac{t}{10}}$.

6. $T = 30 + 60e^{-\frac{t}{20}} + 10e^{-\frac{t}{10}}$, $30\,^{\circ}\text{C}$, 73.86.

Exercises 6.6

1. (a) $y = Ae^{2x} + Be^{-2x}$,   (b) $y = A\cos 3x + B\sin 3x$,

    (c) $y = Ae^{x} + Be^{2x}$,   (d) $y = (Ax + B)e^{x}$.

2. (a) $y = e^{x} + e^{-x}$,   (b) $y = 2e^{2x} - e^{3x}$,

    (c) $y = 2\cos 2x - \sin 2x$,   (d) $y = (5x + 1)e^{-2x}$.

3. (a) $y = Ae^{3x} + Be^{-3x} - \frac{2}{9}x$,

    (b) $y = e^{-\frac{1}{2}x}\left(A\cos\dfrac{\sqrt{3}}{2}x + B\sin\dfrac{\sqrt{3}}{2}x\right) + x^2 - 2x$,

    (c) $y = \frac{1}{2}e^{3x} + \frac{1}{2}e^{-x} - \frac{1}{3}e^{2x}$,   (d) $y = Ae^{2x} + Be^{-x} + \frac{1}{3}xe^{2x}$,

    (e) $y = (Ax + B)e^{x} + \frac{1}{2}x^2 e^{x}$,

    (f) $y = \frac{2}{5}e^{2x} + \frac{3}{5}e^{-3x} - \frac{15}{52}\cos 2x + \frac{3}{52}\sin 2x$,

    (g) $y = Ae^{x} + Be^{-3x} - \frac{1}{6}\cos 3x - \frac{1}{3}\sin 3x$,

    (h) $y = \cos 2x + \frac{1}{2}\sin 2x - \frac{1}{4}x\cos 2x$.

4. $x = \cos\left(\dfrac{t}{2\sqrt{2}}\right)$.

5. $k^2 < \frac{1}{8}$, $x = \left[\cos\{\sqrt{(\frac{1}{8} - k^2)}t\} + \dfrac{k}{\sqrt{(\frac{1}{8} - k^2)}}\sin\{\sqrt{(\frac{1}{8} - k^2)}t\}\right]e^{-kt}$;

    $k^2 = \frac{1}{8}$, $x = (1 + kt)e^{-kt}$;

    $k^2 > \frac{1}{8}$, $x = \frac{1}{2}\left\{\left(1 + \dfrac{k}{\sqrt{(k^2 - \frac{1}{8})}}\right)e^{\sqrt{(k^2 - \frac{1}{8})}t} + \left(1 - \dfrac{k}{\sqrt{(k^2 - \frac{1}{8})}}\right)e^{-\sqrt{(k^2 - \frac{1}{8})}t}\right\}e^{-kt}$.

Exercises 6.7

1. (a) $x = 5e^{-2t}$, $y = (5t + 2)e^{-2t}$,

   (b) $x = 1 + e^{-t} - e^{-2t}$, $y = 2e^{-t} - e^{-2t}$,

   (c) $x = e^{-t} \sin t$, $y = e^{-t}(2 \sin t - \cos t) + 1$,

   (d) $x = 1 - e^{-t} + e^{-3t}$, $y = 3 - 2e^{-t} - e^{-3t}$.

2. $x = (10\ 000 + 49\ 900t)e^{\frac{t}{100}}$, $y = (5\ 000\ 000 + 49\ 900t)e^{\frac{t}{100}}$.

   At $t = 5$, $x = 272\ 805$ and $y = 5\ 518\ 648$.

3. $C_1 = 2.4(1 + 0.25e^{-0.05t})$, $C_2 = 2.4(1 - e^{-0.05t})$.

Index